I0467584

ELECTRIC
BOOM!

THE ULTIMATE GUIDE TO FAST TRACK SUCCESS IN THE BILLION DOLLAR ELECTRIC VEHICLE INDUSTRY

MOHAMMED BELBARAKA

(Usual pen name :SIMON B. BARACK)

ISBN-13: 978-1535072793
ISBN-10: 1535072792

"The true sign of intelligence is not knowledge but imagination."

\- **Albert Einstein**

ELECTRIC BOOM!

CONTENTS

ACKNOWLEDGMENTS

I want to thank my wife and her parents for their support while I wrote this book. This huge undertaking also required the precious help of my friends Courtney, Aladdin and Dominic to which a special thank you goes.

Finally, I want to thank my parents who were powerful role models and who taught me love, kindness and openness.

PREFACE

Congratulations! This book provides you with the breakthrough professional knowledge to grow in the electric vehicle industry.

The automotive industry is changing very fast. Much faster than at any point in the last 30 years. We are now witnessing a significant shift towards green technologies and a more decarbonized world driven by a growing awareness of the general public about the health of our planet. At the same time, people want a better experience with their cars and expect more options like adaptive cruise control, blind spot warning and more connectivity with the outside world while driving.

Consumer adoption of electric vehicles is much faster than that of hybrids, and this is definitely going to accelerate as the charging infrastructure, battery performance, and affordability improves. Every major car manufacturer will expand its EV offering, and soon you will see a hybrid, plug-in hybrid, or full EV variant on every car model on sale.

As technology improves and safety concerns are addressed adequately, self-driving cars may bring disruptive changes in the society with new paradigms for mobility. We might see a shift from the old ownership model to the new use-value paradigm. A self-driving car might even be a source of revenue for its owners in the foreseeable future.

Electric vehicles have become so much more technologically advanced that it's now expanding to areas outside of the automotive field.

Mass market introduction is the challenge. As the EV industry grows, there will be many opportunities for companies who can find ways to reduce the cost of producing the main core components of the electric cars. Expertise in batteries, electric motors, power electronics, advanced controls, and electrical systems will become very highly valued and will generate millions of dollars per year for people, companies and organizations with such expertise.

There is so much information about electric cars out there. To get the basics, you used to have to comb through pages upon pages of complicated and confusing reports, magazines, blogs, and a host of other publications.

Electric BOOM ! is the perfect comprehensive tool to break into the electric vehicle industry, or just satisfy your craving for knowledge about the future.

This book will give you the keys to this industry; it will help you understand how engineers and all the stakeholder community are changing the world.

INTRODUCTION

Everything is about human interactions and exchanges. In a world of more and more complex infrastructure and close to 9 billion inhabitants, the mobility of people and commodities is at the very heart of human species prosperity. In order to maintain and further develop this level of connectivity of the world as we can see today, more and more energy is needed to achieve this goal.

Post industrialization era's infrastructure, based on fossil fuel, is definitely showing its limits both from a geopolitical level and a global warming level. Along the lines of the 1997's adoption of the Kyoto Protocol, in September 2009, both the European Union (EU) and G8 leaders agreed that CO_2 emissions must be cut by 80% by 2050 if atmospheric CO_2 is to stabilize at a quantified number deemed reasonable and keep the global warming at the safe level of 2°C.

Most recently the United Nations conference on climate

change COP21 in Paris ended December 12th, 2015 with the objective of a global warming limitation between 1.5 °C and 2 °C by 2100.

Obviously, one of the biggest areas of action to achieve decarbonization, if not the major one, is the road transport sector. Now, what is the whole industry about? How could I navigate through this massive amount of information and yet be able to digest and retain the most essential part? Like any other subject our brain needs structure, a frame, a pattern, and a context in order to spatially map the subject and connect it with all the existing information, memories, feelings that it has before one could make sense of any information it encounters.

The aim of this book is to take you step by step along the way through the definitions and technical concepts related to the electric vehicle industry in order to help you speed-up your learning curve and increase your ability to dive into much more specific topics while permanently retaining the information and in shorter times. Once the basic concepts are clearly exposed we will focus on the electric vehicle architectures available and explain each of them.

Finally, we will cover in the third and last part of the book, the challenges to mass-market introduction of EVs and the underlying opportunities.

PART 1 – THE BASICS

MOHAMMED BELBARAKA

THE BASICS

This part is about introducing the major technical key concepts to understand how an Electric Vehicle works. You will understand where it came from and where they are headed.

In order to make things easier to apprehend, we will keep alongside the electric Vehicle, the parallel with the conventional vehicle using an Internal combustion engine (ICE) as a means of propulsion. For now this is the most common kind of vehicles to us all – but things are about to change very fast with more and more EVs on the road.

At the end of this part you will be very familiar with the major key concepts for a ground rooted and a firm understanding of the energy conversion's major steps. You will be having enough knowledge not to be impressed by the experts anymore. The underlying challenges will be discussed later in both parts two and three.

For now, we will break down the energy conversion path from the energy source – the fuel tank for conventional vehicle and the battery for an electric car- to the wheels. In this breaking down process we will highlight each of the important stages of energy conversion from the tank to the wheels.

Let's start with very simple concepts that will help you get the big picture.

CHAPTER 1

HOW IT ALL STARTED

Technology and science have drastically improved during the last hundred years of the 20th century. The next five years promises even more amazing developments in the world of technology. One full century ago, the new technologies that had people talking were considered just as groundbreaking. Electricity led the charge of developments that were changing the way people lived every day. It may be hard to believe, but the electric car is nothing new. In fact, the first electric car was introduced more than 100 years ago.

THE BIRTH OF ELECTRIC VEHICLES

It's unclear exactly when the electric car was first invented and in which country. This seems to be more a succession of breakthroughs from the battery to the electric motor that finally led to the first electric vehicle on the road in the 1800s.

In the United States, it was around the end of the 19th century that first electric vehicle made its debut. William Morrison's 6-passengers electric car capable of a top speed of 14 miles per hour helped spark the interest in EVs. Very quickly, by 1900 electric vehicles accounted for around a third of all vehicles on the road. Even New York City had a fleet of more than 60 electric taxis.

As electric vehicles came onto the market, so did the gasoline-powered cars thanks to improvements of the internal combustion engine (ICE) in the 1800s. When first introduced, ICE powered vehicles required a lot of manual effort to drive. In fact, changing gears and manual cranking was not an easy task thus making them difficult to operate. They were also noisy, and their exhaust was unpleasant. This has definitely played in favor of electric cars which didn't have any of the issues associated with an ICE. They were quiet, easy to drive and didn't emit a smelly pollutant like their ICE propelled counterpart.

Thanks to these advantages, electric cars quickly became popular and many innovators at the time took note of this high demand to improve the technology. It's fairly unknown, but the very first Porsche was an electric car - the P1 - in 1898. At 22 years old, Ferdinand Porsche created the world's first hybrid electric car- powered with both an electric and a gas engine.

At the same period also, the world's most prolific inventors Thomas Edison bet on superiority of electric vehicles' technology and worked to build a better battery. One of Thomas Edison's closest friends was another great mind - Henry Ford. They were traveling together, even purchasing neighboring homes. It was not a surprise when the two partnered and put their genius minds together to conceive a low-cost electric car in 1910.

Ford himself confirmed the project in the January 11, 1914 issue of the New York Times:

"Within a year, I hope, we shall begin the manufacture of an electric automobile. I don't like to talk about things which are a year ahead, but I am willing to tell you something of my plans.

The fact is that Mr. Edison and I have been working for some years on an electric automobile which would be cheap and practicable. Cars have been built for experimental purposes, and we are satisfied now that the way is clear to success. The problem so far has been to build a storage battery of light weight which would operate for long distances without recharging. Mr. Edison has been experimenting with such a battery for some time. "

Edison gave an interview to Automobile Topics in May 1914 where he said:

"Mr. Henry Ford is making plans for the tools, special machinery, factory buildings and equipment for the production of this new electric. There is so much special work to be done that no date can be fixed now as to when the new electric can be put on the market. But Mr. Ford is working steadily on the details, and he knows his business so it will not be long.

I believe that ultimately the electric motor will be universally used for trucking in all large cities and that the electric automobile will be the family carriage of the future. All trucking must come to electricity. I am convinced that it will not be long before all the trucking in New York City will be electric. "

Unfortunately, the press seemed to have forgotten about the Edison-Ford project.

Some conspiracy theorists say that the oil cartels went to

Ford and Edison and forced them to abandon the project. As evidence, they point, the "mysterious" fire that nearly destroyed Edison's workshops in West Orange, New Jersey, in December 1914.

The project was abandoned shortly after that due to other priorities demanding Henry Ford's time.

Few months later, Ford's Model T made gasoline-powered cars widely available and affordable at only $650 versus $1,750 for an electric roadster, thus taking more and more market share against EVs.

The very affordable Model T was the beginning of the end of the electric car.

The introduction of electric starters - eliminating the need of manual cranking- and gas becoming cheap and readily available in rural, and remote areas combined with the proliferation of filling stations across the country helped kill and remove EVs for good from the market by the 1930s.

ELECTRIC CAR REVIVAL

Between the late 1960s and early 1970s, rising oil prices and gasoline shortages created a growing interest in lowering the U.S.'s dependence on foreign oil. With the

Arab Oil Embargo in 1973, Congress took note and passed the Electric and Hybrid Vehicle Research, Development, and Demonstration Act of 1976, authorizing the Energy Department to support research and development in electric and hybrid vehicles. Automakers began at this time to explore options for alternative fuel vehicles, including electric cars.

But still, gasoline-powered cars were many folds better than the electric models developed and produced in the 1970s. The performance at the time was very limited, with no more than 45miles per hour and 40 miles of range. It is the growing environmental concern in the 1990s which came with its sets of new regulations that really pushed automakers to improve the EV technology further and come up with more acceptable performance.

In 1996, the most serious electric car to be developed was GM's EV1, a car that was heavily featured in the 2006 documentary Who Killed the Electric Car? Instead of modifying an existing vehicle, GM designed and developed the EV1 from scratch. In 1996, the EV1 was around $35,000. A BMW was $32,900 and as a Mercedes-Benz C280 $35,250. With a range of 80 miles and the ability to accelerate from 0 to 50 miles per hour in just seven seconds, the EV1 quickly gained a lot of public interest.

Unfortunately, the EV1 was never commercially viable because of its high production costs, so GM discontinued it in 2001.

Then, with a booming economy, a growing middle class, and low gas prices in the late 1990s, many consumers didn't worry about electric vehicles and once again, pushing the chance of mass-market introduction of EVs for later. Nevertheless, this time, the big difference with previous starts and stops of the EV industry was that the world witnessed and acknowledged the promise of the technology.

With the worldwide release of the Prius -Toyota's first mass-produced hybrid electric vehicle- a true revival happened. The Prius first went on sale in Japan in 1997 then in 2000 it was subsequently introduced worldwide.

It became an instant success with many celebrities endorsement helping to raise the profile of the car.

Harrison Ford showed up to the 2003 Academy Awards in Prius. Many other celebrities have joined the growing trend. Julianne Moore, Leonardo DiCaprio, Emily Blunt, Cameron Diaz, Demi Moore and many others stars all owned one.

Another event which is by far the most significant and which is pretty much comparable to the Apple success is the emerging of Tesla Motors. Tesla Motors achieved so much so quickly.

They started with a luxury electric car that could go more than 200 miles on a single charge. That was an announcement with a huge echo worldwide. In the short time since the first announcement in 2006, Tesla has consolidated the acclaim for its cars and has become the largest auto-industry employer in California.

This general public receptivity and the vehicle's success spurred many big automakers to accelerate work on their own electric models. The march towards mass-market introduction has now started for good and serious means are now being allocated, both from private and public pockets to further improve the technology.

The demand for electric drive vehicles will continue to climb as prices drop and consumers look for ways to save money at the pump.

CHAPTER 2

A QUICK STEP BACK FOR BETTER INSIGHT

WHAT'S THE ELECTRIC VEHICLE INDUSTRY ALL ABOUT?

Part of the transportation and automotive industry, the electric vehicle industry is all about addressing the mobility needs with vehicles using electricity as their primary source of energy. Mobility is the ability to move physical objects, beings or both from a given departure spot A to a final destination B where they are meant for a certain use or purpose. There are thousands if not millions of different kind of mobility needs. Providing reliable and cost effective solutions for this diversity of needs is actually what the transportation industry is all about. In the illustration below we can see a few of the ground transportation means available and which motivates millions of us wake up in the morning, constantly improving, finding better solutions to cover all the transportation needs and wants of this planet.

Figure 1 – Wide range of transportation type

Now think about a car. It carries you, family, friends and possession over multiple destinations, crossing long distances loaded with heavy things in a fairly short time .

Back until close to the end of the 19th century, horse-drawn carriages were used for transportation pretty much the same way cars are used today. The main difference with today's cars is that the time to cross the same distance is way less with let's say a Toyota Corolla than with an 18th century horse-drawn carriage. What makes this huge difference?

The answer is that the energy leveraged in the case of a modern automobile is way higher than the energy leveraged from the mere horse's muscles, hence the possibility to move heavier weights in reduced amounts of time with new ways of leveraging higher quantities of energy.

Figure 2 – Horse vs Car

If we make the assumption of an average quality horse in good condition, of a breed suitable for riding, and conditioned for overland travel, the maximum distance it could travel is between 20 to 40 miles per day. With a regular sedan today, you can multiply the distance and carried load by ten. Distance could be even more if we consider re-fueling and changing the driver.

The first key word in our journey toward electric vehicle industry demystification is ENERGY. In order to achieve mobility we need to leverage energy. But first let's understand what energy is.

In order to achieve mobility we need to leverage energy.

WHAT DO WE NEED TO ACHIEVE MOBILITY?

Mobility is the ability to move in any given direction of our three dimensional world. And energy is the "thing" we need to accomplish this physical action of powering the car and make it move. What is this "thing" that we call energy? Here is one of the official definitions you could find if you do a research online:

"Measure of the ability of a body or system to do work or produce a change, expressed usually in joules or kilowatt hours (kWh). No activity is possible without energy and its total amount in the universe is fixed. In other words, it cannot be created or destroyed but can only be changed from one type to another. The two basic types of energy are (1) Potential: energy associated with the nature, position, or state (such as chemical energy, electrical energy, nuclear energy). (2) Kinetic: energy associated with motion (such as a moving car or a spinning wheel)."[1]

As stated in the definition above, no activity is possible without energy. No mobility and no transportation industry without the holy grail, energy. You will not be able to drive your car if the tank is empty, nor light your house in case of grid outage. You cannot either perform your daily duties if you have not eaten in three days. Complete the list of such situations from your own experience and you can get the real feel of what is energy and why it is so important. Doing such an exercise will also make you realize that energy is a mysterious and magical thing, a kind of spiritual being with the ability to embody different forms and perpetually converting from one form to another in the

---

[1] http://www.businessdictionary.com/definition/energy.html

process *"of a body or system performing work or producing a change"*, as stated in the above definition. Volumes have already been written, and millions of others more could be added without covering all about the energy subject. It is important at this stage that you get a good feel of the concept of energy and make sure you look at it from other different perspectives than the ones you are used to so far.

Always with our goal in mind of electric vehicle industry demystification, let's get a little bit closer. What is an electric vehicle?

An electric vehicle is a vehicle that is powered with energy in its electric form to set itself into motion and perform like drive you from home to work for example – we will later in the book refer to Driving-cycles. On the other hand, a conventional vehicle uses the energy in its fossil form – diesel or gasoline – in order to perform those same duties. But wait a minute. How is this energy used? And how as a result we are able to set our vehicles into motion to achieve our mobility needs?

In fact, as stated in the definition of energy, setting a vehicle into motion is actually producing a change. A change of state of a vehicle, weighting a mass (m), from being motionless into that same vehicle set into motion along the roads and streets. This motion is a consequence of a multiple conversion sequences or chain from the energy source in its initially stored form – in the fuel tank or battery - all the way to the wheels of a vehicle set into motion. The energy went from its original chemical/electrochemical form – fuel and electricity- into its kinetic[2] form in this whole process.

Ok, now let's break this down to the simplest single isolated sequences and follow the energy conversion steps. We will also take this opportunity to introduce and demystify some key physics concepts that will help you understand better. Remember that there is no rocket science here, rather just a few simple concepts and equations, simply introduced to help you unlock some doors in your mind. In order for any vehicle on wheels to be set into motion, an action must be made to make the wheels spin on the desired direction of motion.

The energy conversion path would be here from the tank all the way to the wheels.

Let's pause for a minute to introduce one of the simplest, yet most revolutionary machine mankind has ever invented: The wheel.

---

[2] Kinetic is energy associated with motion- refer to the energy definition.

MOHAMMED BELBARAKA

CHAPTER 3

DISSEMINATION OF KEY CONCEPTS

WHAT IS A WHEEL, AND WHAT IS SO REVOLUTIONARY ABOUT IT?

We all know what a wheel is, right? We see wheels every day and in different situations of our daily lives, but let's start the chapter with a formal definition quote:

*"A wheel is a circular component that is intended to rotate on an axle bearing. The wheel is one of the main components of the wheel and axle which is one of the six simple machines. Wheels, in conjunction with axles, **allow***

heavy objects to be moved easily *facilitating movement or transportation while supporting a load, or performing labor in machines. Wheels are also used for other purposes, such as a ship's wheel, steering wheel, potter's wheel and flywheel.*

*Common examples are found in transport applications. A **wheel greatly reduces friction by facilitating motion** by rolling together with the use of axles. In order for wheels to rotate, a moment needs to be applied to the wheel about its axis, either by way of gravity, or by the application of another external force or torque."*

Figure 3- Modern Auto-Wheel

The key concept to bear in mind here is that the wheel –
which is spinning around its axle – in carrying the whole
load of the vehicle while keeping the surface of contact
with the floor as small as possible. The whole miracle
about this very old machine, relies in its ability to carry
heavy loads while at the same time reducing drastically the
friction with the ground.

Think of it this way; let's say we need to move a block of
100kg on a flat and asphalted floor across a given distance
between spot A and B. The illustration below is showing
how using wheels will help reduce the friction resistance,
thus making motion much easier.

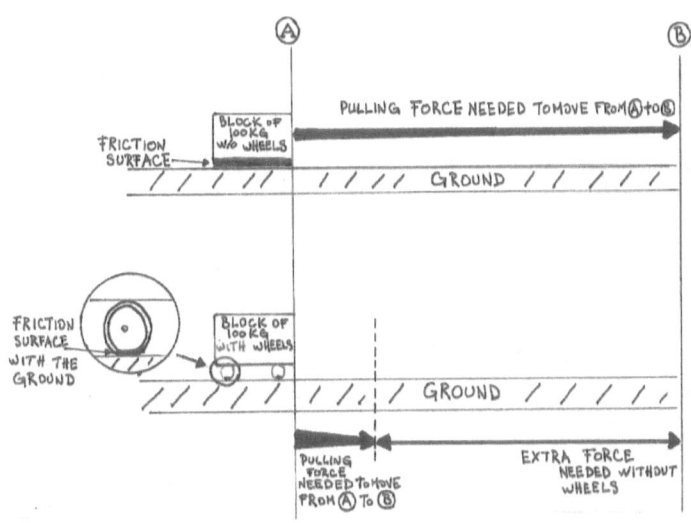

Figure 4 – Magic of the wheels

Without the wheels, the entire surface of the block is in contact with the floor and in order to move the block you need to overcome every Newton (N) of force generated by every single square inch of surface of contact with the floor.

On the other hand, if you put this same block on a cart and even with the new added mass of the cart itself, you will need to apply way less amount of force – Newtons (N)- to move this same block from A to B. We could even say - if we assume the cart weight at 0kg- that the ratio of force to move the block on its wheels with force to move it without wheels is equal to the ratio of contact surface with wheels and contact surface without wheels:

$$Fw/F = Sw/S$$

- Fw = Force in (N) to move the block with wheels
- F = Force in (N) to move the block directly lying on the ground without wheels
- Sw = Surface of contact (m²) with the ground of the block with wheels
- S = Surface of contact (m²) with the ground of the block directly lying on the ground without wheels

---

Once the vehicle/load is on its wheels we just need the

motion power. This used to be- before combustion engine era – horses.

Figure 5 – Horse-drawn carriage

From the beginning of industrial era, the steam engine and combustion engine has been introduced as means of power to replace muscle. Doing so, more energy was now leveraged and more work could be achieved in reduced amount of time.

Horses used to pull the vehicle in order to move it. Now that we removed the horses as means of leveraging energy, how could we achieve the vehicle motion?

It is obvious that if you remove a pulling force from the

front or the back of the vehicle, you need to have someone inside your wheel to spin it, for instance a little mouse or squirrel?

Figure 6 – Spinning the wheel from the inside

Ok you get my point. When you remove the horse, you will all say that you should replace it with an engine, right?

Then what is an engine and how do I connect it to the wheels?

First let us correct a very common misconception among the public, and give the right definition to the reality of things:

> **The engine is merely the means to convert energy and not the energy itself.**

Let's have some fun and learn some more with the next part.

ENGINE DEMYSTIFICATION: WHAT IS AN ENGINE?

Here we are. We introduced energy earlier and its ability to change forms, and while changing from an initial type to the other is changing the state of the objects it is applied to. In our example of a vehicle it is a change from a state of being motionless into a state of motion. The engine will act as the energy conveyor, the energy vector or in other words the channel through which the energy which is stored on the tank – for a conventional vehicle- or on the batteries for an electric vehicle – will flow to change from its original form - chemical[3]- into its kinetic form. The more energy is flowing through the engine, the more speed - kinetic energy- will the vehicle acquire.

Now how does an engine work? And what is the difference between a combustion engine and an electric engine?

Figure 7 – What is under the hood ?

---

[3] Energy stored on fuel or gas, and the one stored on batteries are both chemical energy of different kind. The former is used on combustion engines (ICE) and we rely on the explosion of the fuel when fired to push on pistons and generate the rotation of the crankshaft, the latter is chemical mixtures of ions that generate a flow of electron- electricity- when a difference of potential –load- is applied across the negative and positive poles.

HOW DOES A COMBUSTION ENGINE WORK?

An internal combustion engine, as stated in its name, is using the combustion of the fuel inside the engine's combustion chambers to produce motion. Each explosion is generating pressure inside the combustion chamber which as result will push the pistons. The piston's movement is a vertical back and forth translation inside the cylinders of the combustion chamber as the explosions occurs. The pistons are mechanically linked to a part called crankshaft which as a result of the vertical back and forth movement of the pistons will make the crankshaft rotate. And here we are. From an explosion causing a vertical movement we end-up with the rotation of the crankshaft - because of its specific shape and the "rotational" degree of freedom it has with the pistons.

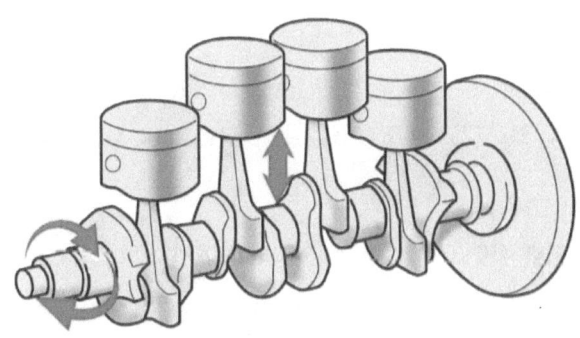

Figure 8 – Pistons & Crankshaft movement

These explosions happen following very specific sequences which help achieve specific rotation speeds and torques at the crankshaft level. A tight control over the oxygen/fuel mixture concentration, quantities and frequencies of the explosion will make it possible to get accurate control over speed and torque levels required at the crankshaft level.

The rotating movement of the crankshaft is then transferred to the wheels through – what we will later discuss- the transmission.

As you will see next, the combustion engine is a very complex product compared to the electric motor.

The relative simplicity of the electric motor over the Internal Combustion Engines (ICE) is opening the doors for new players to challenge the long-established automotive manufacturers with very high expertise in fossil fuel engines.

HOW DOES AN ELECTRIC MOTOR WORK?

An electric motor is composed of a rotor – rotating part connected to the output shaft- and a stator – that static part. It exploits electricity and magnetism to achieve the rotation of the rotor. No explosion and no smoke to evacuate like with ICE, rather just a flow of electrons – electricity-leading to the rotation of the rotor. It would not worth it here to go into the technical details and all the physics principles. There are tons of books and websites available that cover the subject in much more detail.

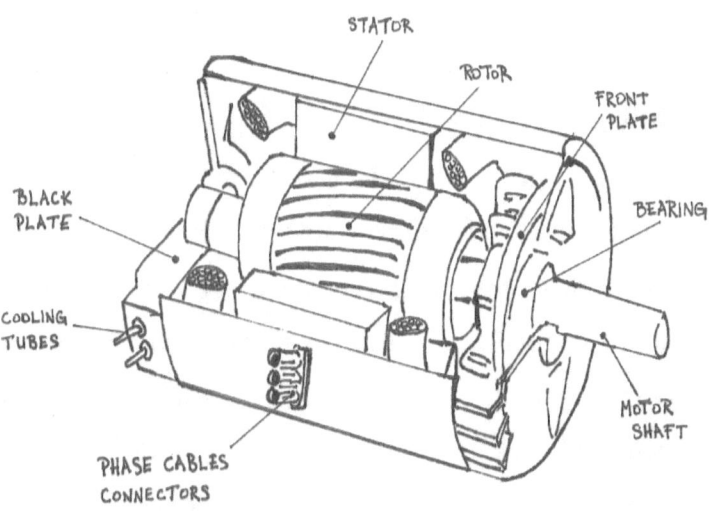

STATOR
ROTOR
FRONT PLATE
BLACK PLATE
BEARING
COOLING TUBES
MOTOR SHAFT
PHASE CABLES CONNECTORS

Figure 9 - Major Components of an Electric Motor

What is important to highlight though is that with an electric motor, energy flows in both directions. In fact it could take energy from the battery and convert it to kinetic energy - embodied in the rotating shaft - or if an external force is applied to the rotor "making it spin", the electric motor would then return this energy back to the battery. In this direction of energy flow we call the electric motor a generator. It is the same machine but it is generating electricity instead of motoring to move the vehicle. It is traction mode of operation versus the generation mode of operation.

Another aspect is that there is always a conversion stage between the battery and the electric motor/generator. In fact a battery is providing direct current (DC) whereas the motors commonly used in the electric vehicle industry for passenger vehicle cars and commercial vehicles mostly needs alternative current (AC) flows to make its rotor spin at the desired speeds. This conversion stage is called an inverter. The inverter is the conversion stage between the battery and the motor and it is converting current from DC- direct current- to AC – Alternative Current- which is generally a sinus current wave. By controlling the amplitude and frequency of this current wave we can control the torque and the speed of the rotor. Don't worry for now if you don't get what torque is. This will be detailed in the next chapter.

The rotating movement of the motor shaft is then transferred to the wheels through the transmission.

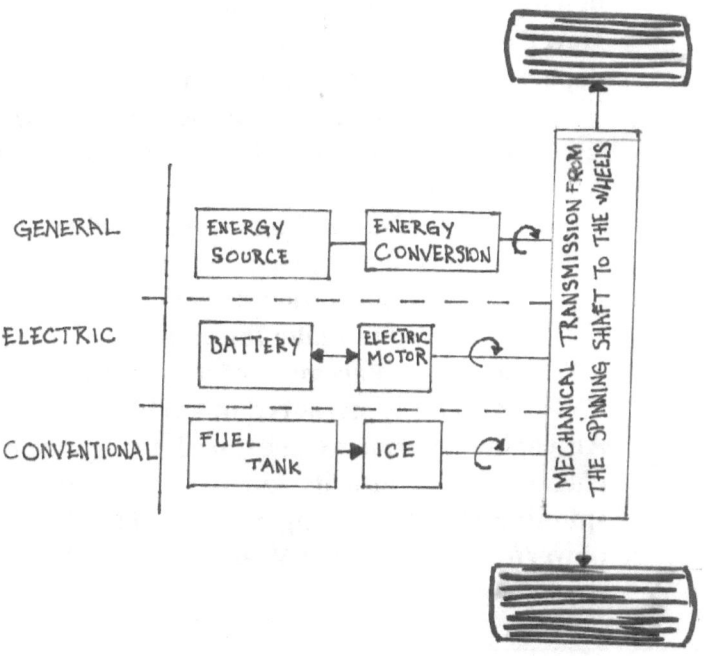

Figure 10- Simplified Energy Conversion Model

The illustration above is summarizing in both cases – combustion and electric- the energy conversion paths from the energy reservoir to the wheels. Now let's summarize the concepts we have introduced so far.

Ideas in brief:

- **A vehicle is a means of mobility it can transport a payload[4] whether it is persons or merchandizes weighing a certain mass from the departure spot A to the destination point B.**
- **How do we transport this payload from departure point A to destination B?**
    1. **We put it on wheels because it will be easier to move with less friction in the ground.**
    2. **We apply a pulling force: A pulling force applied for a given duration of time (t) towards the motion direction, is an energy leveraged to make the wheels spin leading to the displacement of our load from A to B. (This is measured in kWh[5]).**
- **An engine transforms the stored energy from its chemical form into its kinetic form which is a spinning shaft. The bigger the engine, the more powerful it is – more kW or HP- and the more energy it can transfer.**

---

[4] Payload is the max load in kg or lbs that the vehicle is designed to carry. It is a very important specification for commercial vehicles. It is very important for a fleet manager to understand how much their vehicle could carry, since this is reflected in the fleet's direct income potential.

[5] kWh for Kilowatt-hour. One kWh is one kW of power applied on a system during one hour.

Before we move to the next part, let me highlight additional very important information to bear in mind at all times that energy transfers are involved.

If you remember well the energy definition, it is stating that it transforms from a type to another and the quantity of energy is always the same in the universe. But the problem is that in real life and for our industry, not all the energy which is present in the reservoir- in its stored form- is 100% transferred to the vehicle in order to set it into motion. What? You would argue, and I understand, that the definition states that there are no losses. So why can't I have it all? Where does the remainder of it go?

Indeed, the definition is 100% correct, but during the process of energy conversion and because our machines and systems are never perfect, there is always a loss in the conversion processes. One of the most critical points to the engineers in the industry is systems' efficiency. In other words, the ability to design an energy conversion path, from the tank/batteries to the wheels, that makes the difference between the amount of energy in and energy out as small as possible. In real life energy flowing through a path is like a water hose which is leaking all the way from the tap to the end-side of it.

All what we call losses are merely deficits from our

perspective of not having the possibility to use 100% of the reservoir potential to our final need. What actually happens is that the part of what we call energy lost is an energy portion that our system have not been able to capture under the needed and useful form for our purpose, rather this "lost" portion is transformed to heat and friction - causing wear and tear- instead of kinetic energy which is the energy form we want for our purpose of achieving mobility.

The objective of the this book is not to cover the details of energy transfers, rather just highlight the importance of those losses in the industry, since this is the number one constraint on engineers:

Maximize performance and efficiency while reducing costs.

Illustrations below explain the different kinds of losses for each of the conventional vehicle and EV all along the way of the transformation/conversion process from the tank/battery to the wheels:

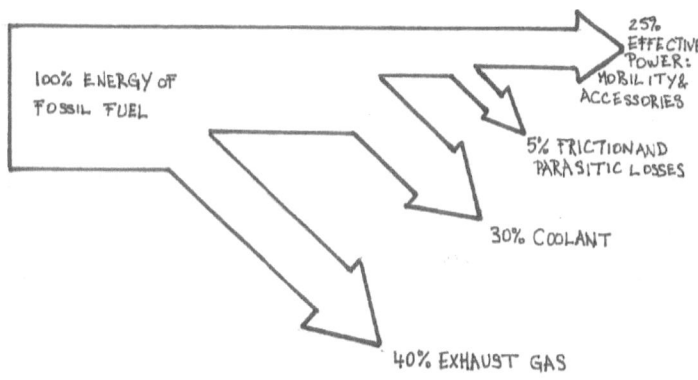

Figure 11 – Energy Split From Fuel Tank to Wheels

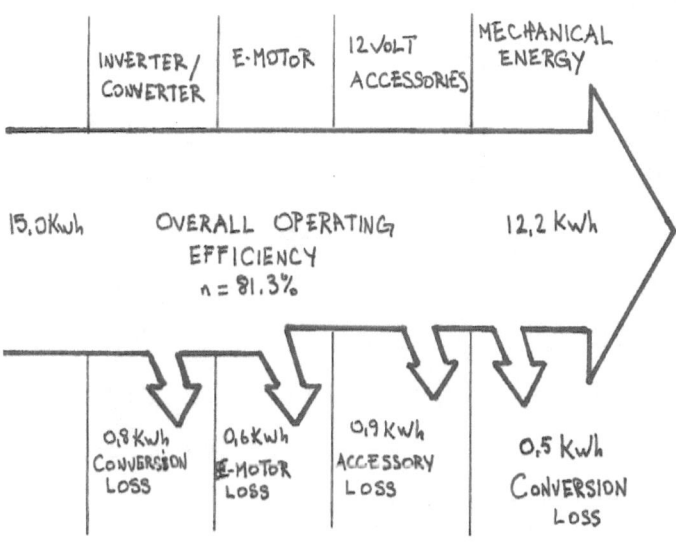

Figure 12 – Energy Losses from a 15kWh Battery

We are now familiar with the most basic concepts to continue our journey safely and not have any confusion about what will follow next.

We'd like to take this opportunity to share a thought. Learning is like eating, it is supposed to be fun and very pleasant all along the way. Imagine an all-you-can-eat buffet full with top chef's meals and treats of all kinds. When you look at it you feel very excited about the prospect of filling your plate and enjoying the taste, don't you? Imagine now that the rules change. You can no longer cherry pick from what you feel like eating or trying, the rule now is that you have to eat it all, and a fine of one hundred bucks for each left over piece will be charged to you. What a nightmare! The food is still awesome right? But just a change in the rules makes the whole experience from being very pleasant to very ugly and painful. This is the same about knowledge. Addressed and taught the proper way will make the journey very pleasant and stimulating.

Hope you are still enjoying the ride. There is a whole bunch of very exciting stuff to come about this very exciting thing: the electric vehicle industry.

You will soon gain as much insight as an insider of the EV industry.

ELECTRIC BOOM!

QUANTIFYING THE ENERGY AND ITS FLOWS - INTRODUCING TORQUE, SPEED AND POWER CONCEPTS

In order to dive a little deeper, we want you to be familiar with a few more new vocabulary terms and concepts in order to be more comfortable and to gain a deeper understanding. You will then be able to have some smart discussions with colleagues, friends, and even with experts of the industry! From now on, you will be able to raise your insight in this field and not be impressed anymore by technical guys.

Remember that the whole purpose of electric vehicle industry is to provide a means of mobility to transport payloads, persons, merchandises and commodities of all kinds while being environmental friendly and not harming the planet. The best way to achieve this movement is to put wheels under these loads and use electric energy – or to be more accurate, have this electric energy change from its stored type to its form producing a change of state to the vehicle[6]- to drive or spin your wheels in order to move.

Now, during all driving conditions, you know that the

---

[6] Remember Energy Definition.

journey from the departure spots to the final destinations is never a straight line on a flat, clean and empty road. The journey from departure to destination, from A to B, for pretty much all kinds of ground transportation is plenty of stops, starts, accelerations, decelerations, climbing hills, downhill slopes, pedestrians, other cars and bicycles to watch, and all sorts of other different obstacles. This means one needs to have a very tight control over energy conversion to perform the A to B journey which we will call from now on the drive cycle or road profile. Below, an illustration of what could be a typical urban drive cycle:

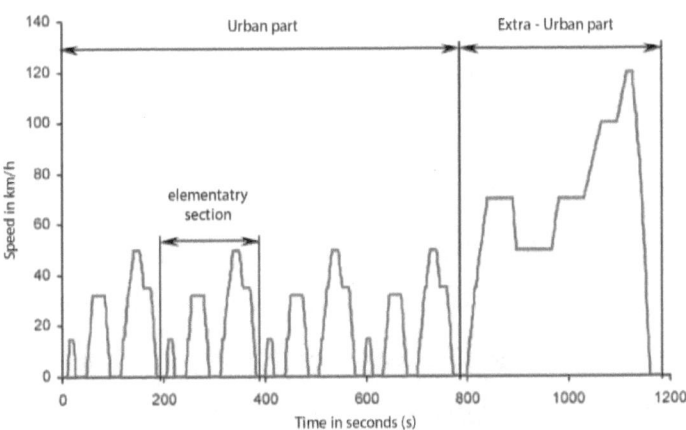

Figure 13- New European Driving Cycle, NEDC.

What we see in the graphic above is actually a

standardized driving cycle defined by government institutions- in the case the World Forum for Hamonization of Vehicle Regulation[7] - in order for OEMs (Original Equipment Manufacturer) to test their vehicle against to certify both performance and emissions' requirements set by those same agencies. It is a driving cycle of 1200 seconds' duration – 20 minutes- where the vehicle has to follow a certain speed profile. If we look at it from the beginning – left bottom corner of the graph- and imagine a vehicle performing the drive cycle, we will see it accelerate from 0 to close to 20km/h following the ramp in the graph, than maintaining that speed for couple of seconds, decelerate to 0, stop for another couple of seconds, reaccelerate following the ramp to 30km/h, maintain the speed for another couple of seconds etc... until the end of the cycle. This kind of cycle is not necessarily representative of an actual real life driving situation- what we mean here is that there is maybe no real life road where you could replicate exactly this driving cycle- rather it is a compilation of different urban/suburban driving conditions in order to reproduce a condensate of all different conditions in a single testing set-up. This is generally

---

[7] This is a working party of the Inland Transport Division of the United Nations Economic Commission for Europe.

performed on a chassis-dynamometer which helps reduce both tests costs and duration.

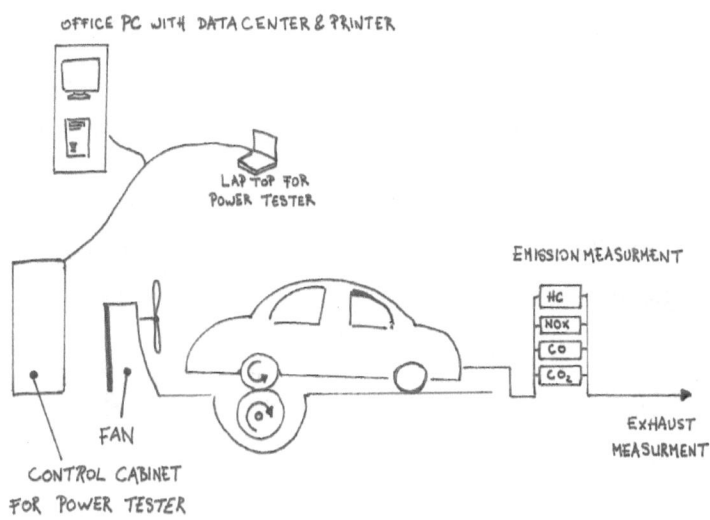

Figure 14 – Vehicle running on a Chassis Dynamometer

Important Fact:

A big portion of the efforts and before the introduction of any product into the market is testing. Testing and validation happens at all levels from the simplest components like nuts and bolts, to subsystems, all the way to much more complex systems and vehicle levels. It is a very structured area of the industry and represents an industry for itself with large organizations both private and public specializing in specific testing areas.

For each step of the driving cycle a specific amount of energy is needed, even more, this specific amount of energy needs to be dispensed following a specific rate and at specific timing and certainly not randomly. Let's zoom-in a typical section of the duty cycle and see how the energy is flowing out of the reservoir[8].

---

[8] In fact energy flows happens in both directions in an electric vehicle. The energy flows out of the battery and back in again in some driving situations (regenerative braking).

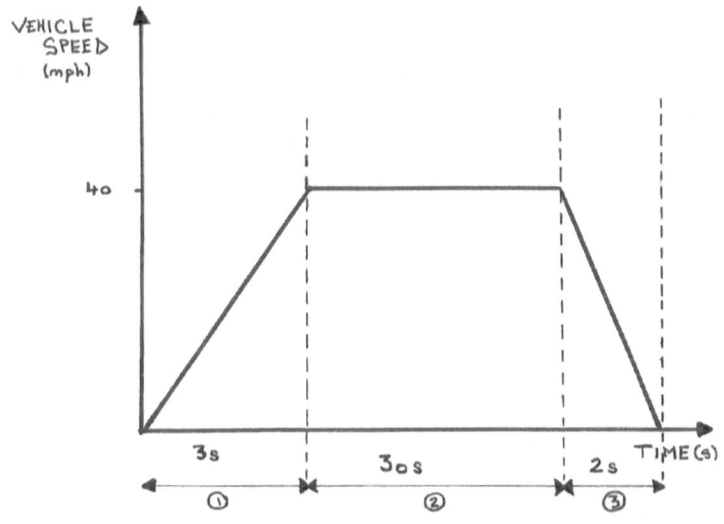

Figure 15- Simple Driving Cycle Sequence

The section in the figure above could be a situation for example where a vehicle is at stop and needs to reach the next street corner which is a few blocks away in some 300 metres' distance. Before the vehicle reaches to the targeted street corner, the traffic light turns red which leads the driver to stop.

Ok let's break the complete small sequence down to more simple sections in order to highlight what happens from a system change of state perspective[9] – what I call system here is the vehicle. The vehicle is at stop then it accelerates

---

[9] Remember the definition of Energy

from 0 to 40mph (miles per hour) in 3s (1st section), then maintains this 40mph during 30s (2nd section) and finally decelerates from 40mph to 0 in 2s (3rd and last section).

At each of those phases there are energy transfers. The first transfer of energy happens when the energy stored in the vehicle - either fuel on the tank or current on the battery- is transferred to the motor shaft which- through the transmission- make the wheels spin. We will see in the next chapter how the motor shaft is connected to the wheels and how a transmission works.

This energy is transferred at the specific rate and the specific quantity that is needed to make it possible for a vehicle of that size and mass to reach 40mph from standstill in 3s. Always keep in mind the idea that a vehicle is a load or mass. To move this load you need an energy transfer to change its states of motionless into motion. The velocity (speed) of the vehicle is actually an image of energy playing live in its kinetic form. We are using purposely odd words here which are absolutely not scientific in order to make you feel this energy transfer's concept. What is fun here is that once you are comfortable with this concept for the vehicle, you can understand any other kind of energy transfers. This will be very helpful in the next chapters to understand the different Hybrid Electric Vehicle (HEV) architectures and the articulation between EVs, renewable energies and smart grids. It is all about what type of energy, who needs it, when, where, and in how much quantity. The whole game is about energy transfers. We are familiar with the adage "time is money." You will see in next chapters that "energy is money" is not less true.

So if we return back to our small section's description, we can see that a certain amount/quantity of energy needs to be transferred from the reservoir to the wheels in order for the vehicle to go from 0mph – motionless state - to 40mph in 3s. During these three seconds the wheels needs to spin at faster and faster pace while carrying the vehicle load until it reach the spinning speed – rpm[10] rate- that makes it possible for the vehicle to reach the 40mph of speed desired.

In order for the vehicle to accelerate following the ramp in the illustration above, a certain amount of torque has to be applied to the axle of the wheels – the amount of torque that is needed to achieve the 0 to 40mph in 3s for the given vehicle weight. For a given acceleration profile or ramp,

---

[10] rpm – revolution per minute

this amount of torque needed will vary as a function of the vehicle load. And the more the vehicle is heavy, the more torque is needed to achieve this same acceleration. Likewise in case of an uphill section, the vehicle will need more torque than in a flat one. See below:

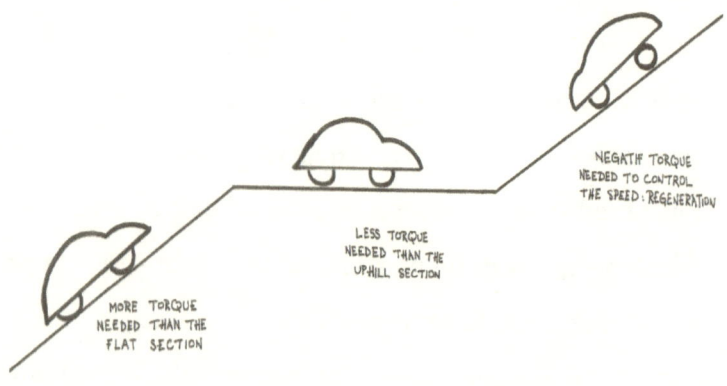

Figure 16 - Example of a road profile section

I am still confused, what is torque? Don't worry, below a few concepts are introduced to help understand all the physical magnitudes involved in the process of transforming/transferring energy.

- **From rpm of the wheel to vehicle speed:**

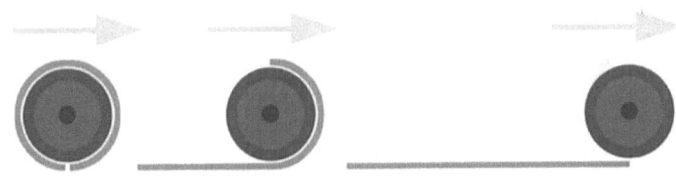

Figure 17 – Distance = (circumference) X (number of revolutions)

The perimeter of a circle is more commonly known as the circumference. Its value is :

Perimeter =2 . π . r

(r=radius of the wheel and π=3.14)

In the international measurement system it is expressed in meters.

As you can easily see from the illustration above, each time the wheel revolves around its axle, the vehicle moves in translation by the equivalent distance of the wheel circumference. If the wheel circumference of our car is 2.13 m and the wheels are revving at 1000 rpm (revolution per minutes), this means that the vehicle will move by 2130 metres (2.13 x 1000) each minute, this is 2.13km/min which is: 2.13x60 = 130 km per hour.

Now, and please give us all your attention here because the answer is about to come for what is torque. In order to make the wheels spin at the desired rpm rate- and don't forget that it is a wheel of a vehicle which is carrying a certain load - we need to apply an appropriate torque to the axle of the wheel to obtain the desired rpm rate or in other words to make it spin at the desired speed.

Look at the illustration below, this should be now clear:

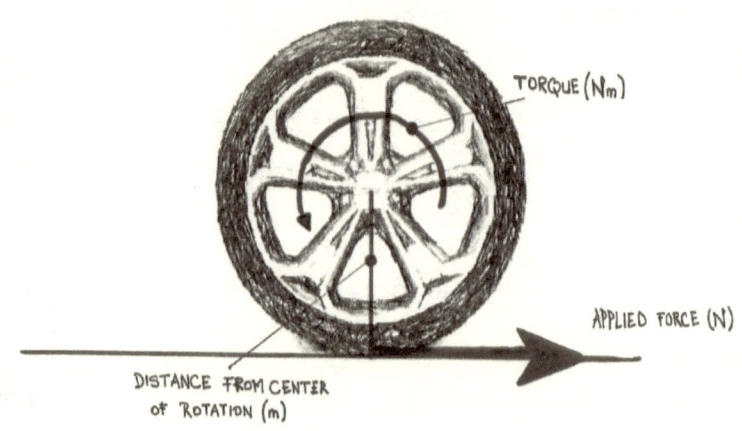

TORQUE (Nm) = FORCE (N) x WHEEL RADIUS (m)

Figure 18- Relation between Traction Force and Torque

The torque is the equivalent of the tractive force F times the wheel radius r :

Torque (Nm) = Wheel radius (r in meters) . Applied Force (N)

or

$$\underline{T=r \cdot F}$$

Torque is expressed in Nm (newton meters). This torque is obtain either by applying an external pulling force (F) which was used to be applied by the horse drawing the carriage or- in the case of a modern car- directly to the axle by the engine's shaft. The engine's shaft is applying torque directly to the wheels through the transmission gears that we will detail in the next chapter.

Ok, you have your wheels carrying the load of the vehicle, which need a certain amount of torque in order to reach a certain rpm rate and thus for a car to reach a certain speed. You see here a direct relation between torque and RPM rate – rotation speed – of the wheel. This relation is what we call power. Let's have a look.

- What is Power?

Power is a physical concept or tool used to simplify the description of physical situations or behaviours, whether electrical or mechanical, when two physical parameters influence the behaviour of a system. In our example of a vehicle[11], the power is a function of both Torque and speed P(T,speed).

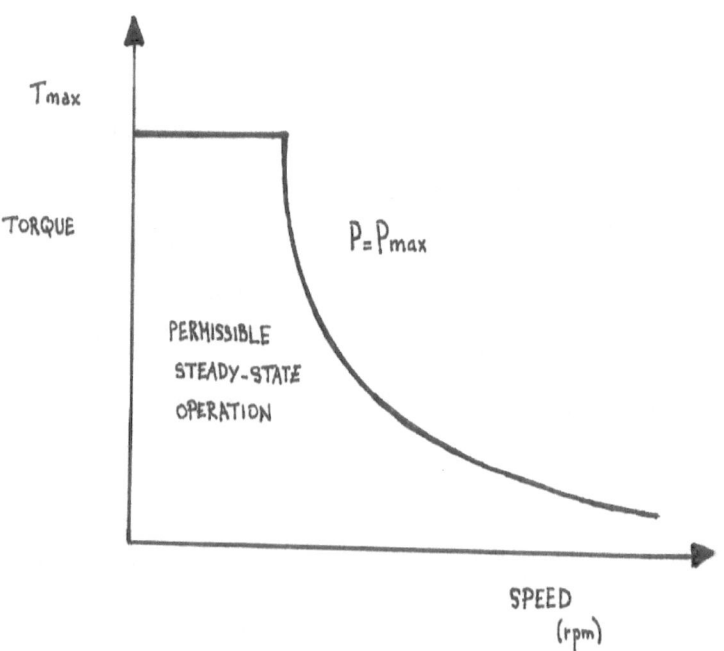

Figure 19 - Torque vs Speed Curve of an e-motor

---

[11] For a battery, the power is a function of Voltage and Current-Pbat=V (voltage in Volts -V) x I (current in Amperes- A).

When the vehicle is driving on a road- if we look at the driving cycle curve which is vehicle speed as function of time- there is an equivalent amount of torque applied to the axle of the wheels at each time the vehicle is in motion. The power in Watt (W)[12] is the result of Torque (Nm) multiplied by the angular speed of the wheel in radian per second (rad/s):

Power (W) =Torque (Nm) X Speed (rad/s)[13]

This means that at each time of the driving cycle an equivalent amount of power is requested to be able to follow the drive cycle. The next curve is the representation of the power requested at each time in order for a regular sedan of 1500kg to be able to follow the so-called NEDC driving cycle.

---

[12] Watt is the expression for power magnitude (W).
[13] Each one rpm = 0.1046 rad/s (1rpm= 2π rad/60s = 0.1046rad/s).

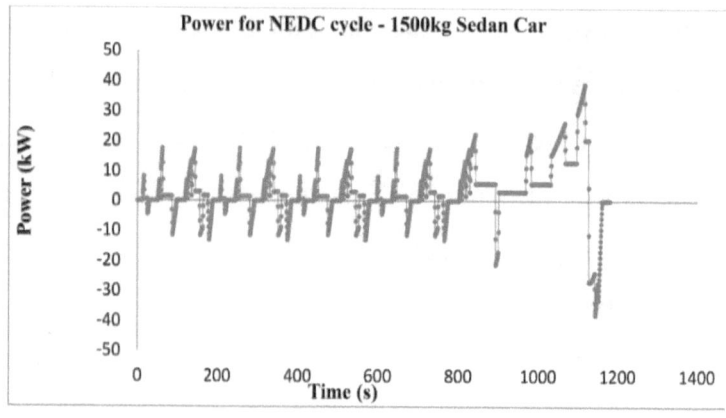

Figure 20 – Power NEDC cycle for 1500kg car

This graph shows exactly how much power a 1500kg vehicle needs at each time during the NEDC driving cycle. If you look at the graph you will notice that there is some sections during which the amount of power request is negative. How come? Actually negative power only means that the vehicle is decelerating. This corresponds exactly to the sections of the driving cycle requesting the vehicle to decelerate. Next figure highlights some of the deceleration areas of the cycle:

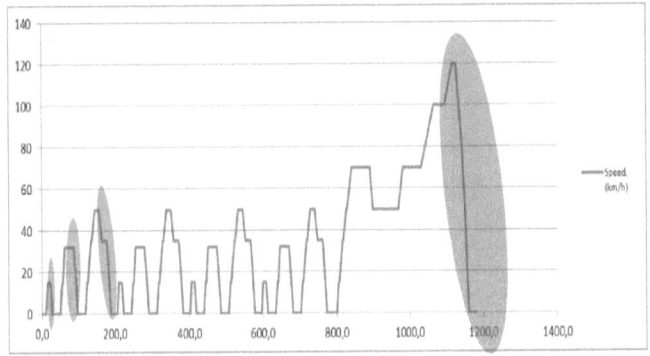

Figure 21 – Highlight of some deceleration section in the NEDC Drive Cycle.

Deceleration means that kinetic energy of the car needs to be absorbed somewhere else in order for the car to stop, this is either absorbed by the hydraulic braking system of the car – which applies a force on the wheels' brake discs and thus transforming the kinetic energy of the car into heat[14]- or - and this is what makes electric cars interesting as opposed to ICE propelled cars- is to reverse the torque direction on the electric motor shaft which will generate braking. This is what is referred to in the industry as regenerative braking or regen. In doing so and if the battery is able to absorb that amount of braking power, we are now

---

[14] In fact for sake of simplicity because this is not our subject to cover in this book we only considered here the transfer from kinetic into heat.

able to recover a decent amount of energy for additional range[15].

- How energy links with Torque speed and Power

The last important concept to teach here is the link between torque, speed and power to energy. We are very close! Now that you understand what is torque and speed, and how they relate to power - P= Torque x Speed - energy is merely the amount of power drawn for a time duration. It is expressed in kilowatt hour (kWh). For example, if you draw 20kW of power from your battery during one hour, the total amount of energy you would have drawn is 20kWh. For our specific driving cycle example above, if we integrate P(t) between 0 to 1200s - duration of the cycle- we will get the total amount of energy requested for our 1500kg sedan to perform this drive cycle. Roughly it is 6kW of mean power which is requested during the whole driving cycle - 20 min (1/3 of an hour) - which is 2kWh of energy to perform this driving cycle[16]. And this is the total

---

[15] Vehicle range is the number of km an electric vehicle is able to drive on a single charge of the battery (refer to the glossary).
[16] If we consider the stops durations and assume there is regen braking during the deceleration phases, the real consumption is close to 1kWh for this 20min cycle and 11 km of total distance crossed.

amount of energy required for a 1500kg car to follow the speed profile of the above NEDC driving cycle.

Energy (kWh)= Power (kW) . Time (h)

Energy is the amount of power applied for a given duration of time. You cannot mention energy without associating the time duration under which this power is applied to a system.

One last technical notion on how we make the physical link between an engine and the wheels : The transmission.

Transmission- From Engine Shaft to the Wheels

We introduced the concepts to express and quantify power and energy flows involved in the process of operating and driving a vehicle. Let's now zoom into the section between the motor and the wheels and see how energy is transferred between them.

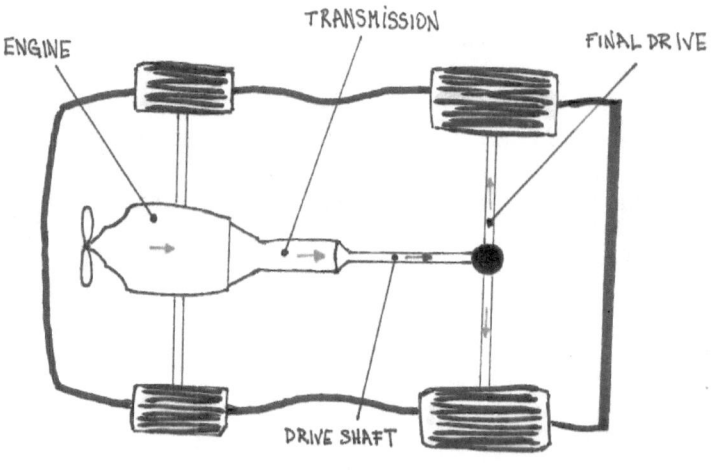

Figure 22 - Transmission linking engine to the wheels

The illustration above shows the section between the motor and the wheels. It is a compilation of rotating parts connected to each other all the way between the rotating engine/motor shaft and the wheels in order to transfer the power from the engine to each of the wheels. This power, as seen in the previous section, is the result of the torque (Nm) times the speed (rpm) expressed in radians per second (rad/s):

$$P (W) = T (Nm) . Speed (rpm) \times 2\pi/60$$

Since we need to achieve specific driving cycles with the vehicles – which require specific rates and quantities of energy transfers, hence determined amounts of torque and speed at each instant of the duty cycles- our transmission line - or so to speak the energy transfer channel - needs to be specifically designed to channel those specific amounts of torques and speeds. For example, if you use a straw to water your back yard you will not be able to do it in a reasonable amount of time - if by any chance the straw doesn't explode under the pressure of water. This is the same for the gear sets arrangement. For torque and speed transfers – basically the power transfer- it needs to have the ability to carry the appropriate amount of torque while being able to rev at the appropriate revolution per minute (rpm) rates required for the given vehicle platform to perform the driving conditions.

Using a combination of appropriate gear sets designs is the technical solution to transfer this power properly from the motor shaft all the way to the wheels of our vehicle.

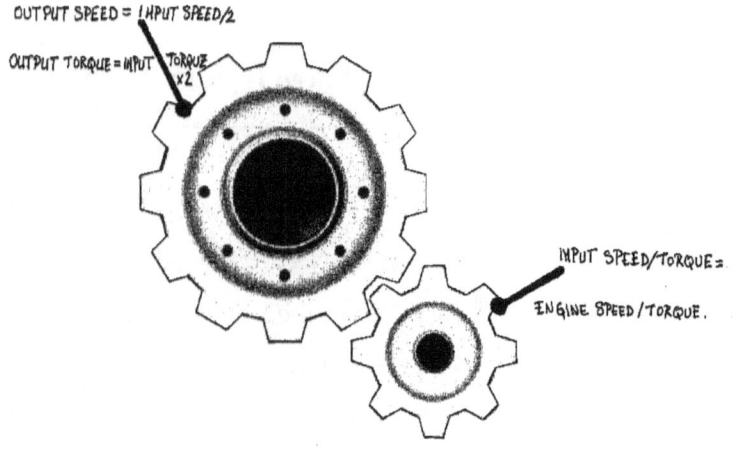

Figure 23 - Gear Wheels of 2:1 ratio

In this simple gear arrangement above, there is a small wheel and a larger one. The smaller gear is half the size of the bigger one. If you connect the motor shaft to the smaller gear –input gear - each time the motor shaft will make two revolutions the bigger gear – output gear- would make only one turn because of the 2:1 ratio between the two gear sizes and number of teeth. Similarly, and since the power remain constant - with Power =Torque x Speed, there is no power creation in the process of one wheel entraining the other- when the output speed is half of the input speed, then the output torque must be double of the input torque in order to keep the power constant:

$$P = T_{Input} \cdot Speed_{Input} = T_{Output} \cdot Speed_{Output} = Constant$$

So we have,

$$T_{Input} \cdot S_{Input} = T_{Output} \cdot S_{Output}$$

$$T_{Output} = T_{Input} \cdot Speed_{Input} / S_{Output} \quad (1)$$

So if input speed is 2 X output speed, (1) is now equivalent to :

$$T_{Output} = T_{Input} \cdot (2. S_{Output}) / S_{Output} \text{ and finally}$$

$$\underline{T_{Output} = 2 \cdot T_{Input}}$$

So when passing power from one gear to the other you increase the speed and reduce the torque by the same proportion or the opposite increase torque and reduce the speed depending which is the input and which one is the output. The flexibility to adjust through the appropriate gear set the torque and speed at the wheel level, gives an interesting degree of freedom for the design of the motor itself. In fact, an electric motor could be designed to rev at higher speeds in order to maximize its power output and packaging – the higher the speed of an electric motor, the smaller the packaging, the lower the cost.

An appropriate gear set design – which is our power transfer channel from the motor to the wheel- will make it possible to adjust the level of torque and speed needed at the wheel in order to achieve the requested driving conditions for our vehicle.

There are thousands of different gearboxes, and gear arrangement designs. Each of these gears arrangement is specific to the needs and constrains of the various applications it is addressing, but in 100% of the cases its vital function is to connect and transfer the power from the engine to the wheels at the appropriate speed and torque levels requested for the application.

For most of vehicle architectures – with the exception of wheel-motor kind of architecture[17]- the axle of the wheel is never in line with the engine shaft. It is always in odd positions that make it impossible to connect them directly in a straight line. One of the many advantages of proper

---

[17] Wheel motors are motors embedded inside the wheels. Along with carrying the whole vehicle load, it is acting as a direct drive to the wheels.

gear-set designs is the possibility to connect two bearing axles which are not spatially in-line with each other and thus transferring the rotation from one axle to the other in different directions and angles. It also provides solutions to get from one single input shaft two output driving shafts in order to drive both wheels of the vehicle- left and right- with a single motor:

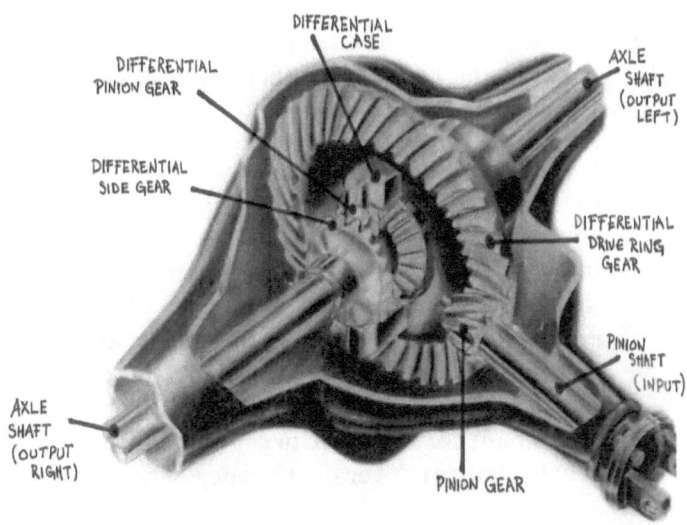

Figure 24 - Inside a differential

Another important advantage of the gear sets design is to operate – when possible – the engine at its best efficiency spots possible by selecting the appropriate reduction ratio (1st gear, 2nd, 3rd...) this is what happens when we shift gears in a manual transmission – automatic transmission are doing the same- we just put the motor in a much comfortable operating condition and generally more efficient while achieving the desired driving profile using the appropriate reduction ratio.

Idea in brief:

The transmission is a set of different gears of different sizes and shapes in contact with each other and specifically designed to connect the motor shaft to the wheels:

- **It transfers power from the engine to the wheels[18]. In this process a reduction/multiplication ratio[19] is always involved.**
- **It makes it possible with specific gear teeth and shapes designs to connect an input to one or multiple outputs in different directions and angles in order to reach the final wheel to drive.**
- **It also makes it possible to reverse the rotation direction from clockwise to counter clockwise.**
- **It helps improve overall efficiency of the system.**

---

[18] This section of the energy transfer path is also subject to losses. In reality only 95 to 97% of the power from motor shaft output is transferred to the wheels.

[19] When you hear saying «shifting gears» this refers to the change in the transmission ratio in order to adapt the engine speed to the wheel speed according to the driving conditions.

This area is an industry itself with its specialized players like GNK, BorgWarner, Getrag and many others. The challenge for this industry is to manufacture gearboxes that have the appropriate geometries and able to transfer the expected amount of torque at the rated speeds while being durable, efficient and cost effective. Development of such transmissions for transport applications cost millions of dollars.

PART 2 – BRIDGING THE GAP - FROM GASOLINE TO ELECTRICITY

ELECTRIC BOOM!

BRIDGING THE GAP

In the EV industry, what we generally refer to as vehicle architecture is how all the components in the vehicle, which are each performing a number of functions, are linked and communicate together in order for the vehicle to perform the driving cycles in the most optimized manner.

The core of this architecture is of course the traction system; the powertrain and this of course includes the battery. As we have seen in the last chapters, the way to achieve traction is to drive the wheels and make them spin at the desired torque and speed rates using the appropriate quantity of energy stored in the tank or in the batteries. It is not always convenient to address all the requirements of an application with a full electric solution where 100% of the energy needed is stored in a battery pack. It always comes down to what is the acceptable cost to the market for such or such application[20]. The cost here has a broader

---

[20] By application here we mean a given transportation need for which we need to design a vehicle.

meaning. It is usually referred to as the Total Cost of Ownership (TCO). The TCO includes everything a vehicle's user and/or fleet manager needs to pay throughout lifetime of the vehicle. This includes, but is not limited to, acquisition price, maintenance costs, fuel/electricity costs, interest, insurance, plus any other expenses during the expected lifetime of the vehicle which is expressed generally in the total number of miles or kilometers.

In fact, given the current technology maturity level, it is still expensive to produce battery packs with expected specifications - powerful, energy dense, safe and which has an acceptable lifespan. Not to mention the volume and weight of current battery packs and all other fairly expensive components like chargers[21], DC/DC[22] converters, and transmissions specifically adapted to electric motors – with higher rpm rates and torque levels to withstand. Besides the cost, weight is a major concern to address as well because, the heavier is the vehicle itself and

---

[21] The charger is the conversion electrical device that will convert the energy from the grid (AC) to charge the battery (DC).
[22] DC/DC is a converter that converts DC power from the battery pack which is generally at a high voltage (350-450VDC) to a 12 VDC output in order to feed devices like fans, vehicle cluster instrument panel, pumps etc. which are the second energy consumers after the traction system.

the less payload to carry, hence less profit particularly for commercial vehicle applications which are very sensitive to this aspect.

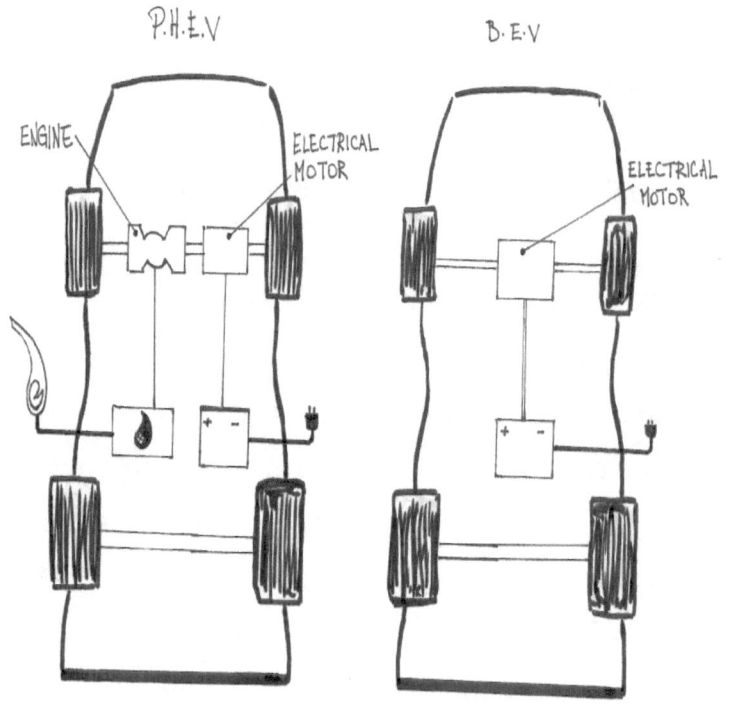

Figure 25– Example of EV architecture.

So since it is not always convenient to store 100% of the energy needed in a battery pack as it is in a fuel tank, and since combustion engines has become more efficient than they used to be back in the 1990s, automakers have come up with different trade-offs depending on the application and their level of R&D and production capabilities. As a result of these trades-offs, different hybridization levels and different hybrid architectures that reflect the technological maturity of the time are proposed.

In fact energy density of fuel compared to whatever battery chemistry type is many folds higher. The batteries are still the least energy dense among any types of fossil fuel. There is nine times more energy per liter of gasoline than the usable energy[23] of a lithium-ion battery for example. It becomes easy understand the difficulty to compete with a diesel or gasoline fuel tank in term of additional weight and space in the vehicle, not to mention the additional cost involved.

One way of overcoming these challenges is hybridization, by mixing two different types of energy: electric and fossil fuel. In fact, Internal Combustion Engines (ICE) and electric motors have different behaviours in terms of performance- torque, speed and efficiency. An electric motor can output its maximum torque right from zero rpm. It does not need to idle at stop and is much more efficient

---

[23] It is 20%-80% State Of Charge. In order to maximize the lifespan of a lithium-ion battery it is recommended to not charge it until 100% and not deplete until 0%. The 20%-80% window of SOC is the comfort zone that maximizes battery lifespan.

than an ICE. On average an electric drivetrain has 80% efficiency from battery to wheels compared to 25-30% for the ICE. An electric motor could regenerate energy back to the battery whereas energy flows in only one single direction with an ICE. An ICE burns too much fuel in some specific areas of the driving cycle as opposed to an electric motor that can operate at higher levels of efficiency on these same spots of the driving cycles and is able of recovering energy back to the battery in some areas of the duty cycle. This makes both engine types complementary and using them both in a vehicle leads to significant fuel savings, hence reduced environmental footprint and sometimes reduced Total cost of Ownership (TCO).

An additional source of energy associated with a good energy management strategy, helps make the appropriate decision as to where to draw the energy from and when according to the component's limits and driving conditions.

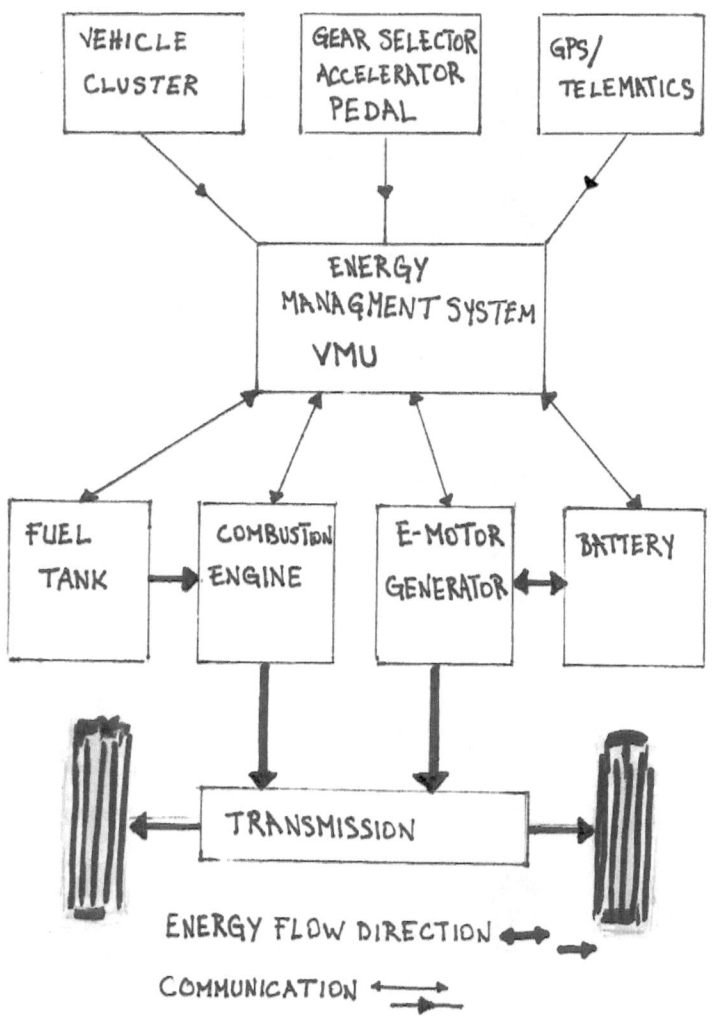

Figure 26 – Energy Management: Control Structure

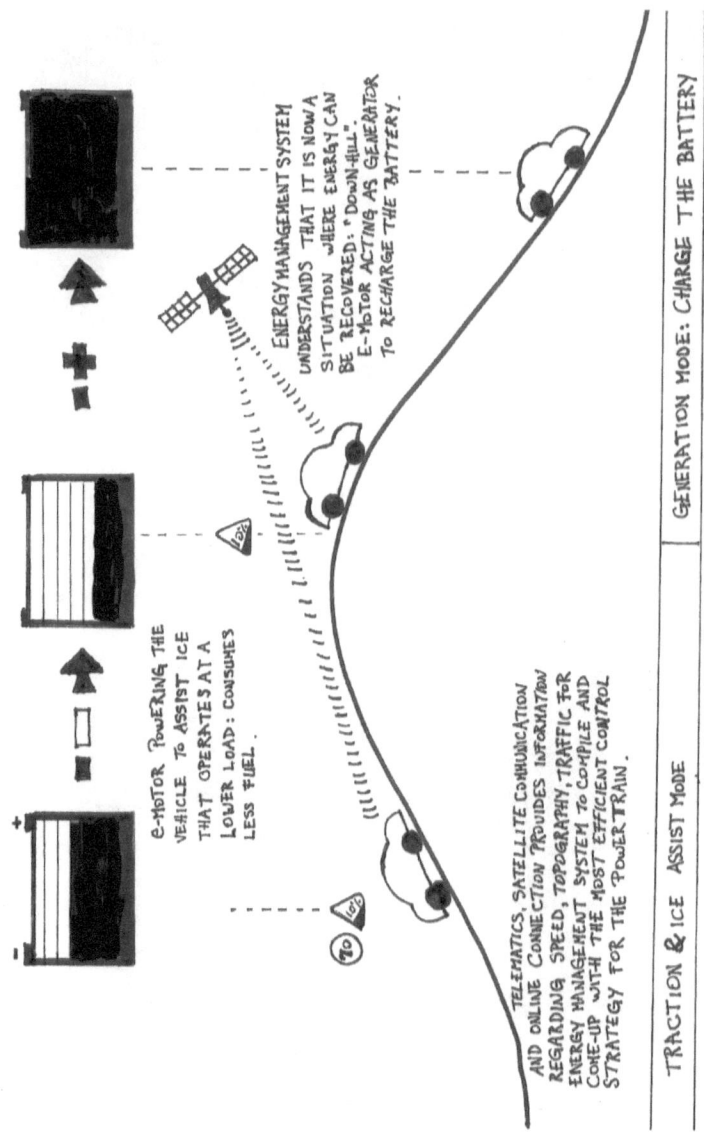

Figure 27 - Energy Management: Example of an HEV

Idea in brief:

Whenever we talk hybrid, it is at least two forms of energy stored in the vehicle. How these energies are used, when and for what purpose will differentiate between the multitudes of hybrid architectures available. We can classify the hybrid architectures in two categories: Series Hybrids and Parallel Hybrids. The difference between those relies on the connection of the ICE to the wheels or not.

In the series hybrid only the electric motor is driving the wheels, whereas the combustion engine is coupled to a generator and used merely as a gen-set to charge the battery.

In the parallel hybrid architecture both engines electric and ICE could drive the wheels. Depending on the architectures, there could be different operating modes :

- Full electric
- ICE only
- Both at the same time driving the wheels

CHAPTER 4

ELECTRIFICATION LEVELS

We will discover now the most common hybrid configurations known to the industry with potential market outputs.

FULL EV

A Full EV or a battery electric vehicle (BEV) is a vehicle that is powered entirely on electric energy; this necessitates an electric motor and battery pack that are respectively powerful enough and large enough in order to perform the driving cycles. Just like Tesla for example.

There are many BEV design variations based on the e-motor size –max torque and max rpm- and transmission line design; use of a clutch, differential[24], multispeed gearbox. However, a full EV architecture will always work according to the schematic below.

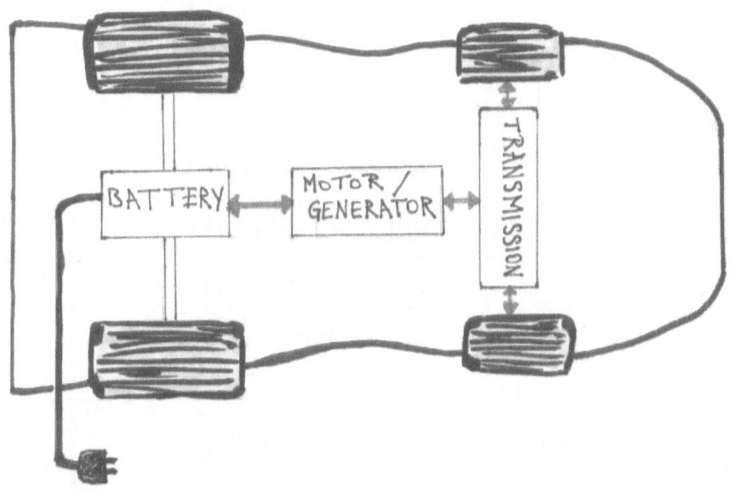

Figure 28 – Battery Electric Vehicle Powertrain

---

[24] The differential is the device that splits engine torque two ways, allowing at the same time each output to spin at different speed (when the vehicle is turning or taking a curved route the wheels at the outside of the curve must turn at higher speed than the wheels inside the curve).

As introduced in the beginning of this part of the book, full EV is facing two major challenges that slow mass market penetration. The first is the cost of the battery pack which makes the vehicle much more expensive than the conventional powertrain, and the second is the availability of charging infrastructure which restrict the use of BEVs to urban areas equipped with charging infrastructure. Besides the cost, packaging and weight of the batteries, BEV architecture is definitely the easiest architecture to implement because of the simplicity of integration and development, and stabilization of the controls.

MILD HYBRID ELECTRIC VEHICLES

As specified earlier, a hybrid electric vehicle (HEV) relies on two energy sources, usually fuel – which is converted to mechanical energy thanks to an internal combustion engine (ICE) - and a battery using a motor/generator for the energy conversion. What is referred to in the industry as a Mild Hybrids are the least electrified type of HEVs. A Mild Hybrid is a conventional internal combustion engine (ICE) vehicle with a larger starter motor that can also be used as a generator, usually called an integrated starter-generator (ISG), and a battery – larger than the usual 12V lead-acid batteries in conventional vehicles- that is powered and is recharged by the motor in the generation mode. In this kind of architecture, the ICE is still the main traction means and must always be on while the vehicle is moving. You would argue, what are the benefits?

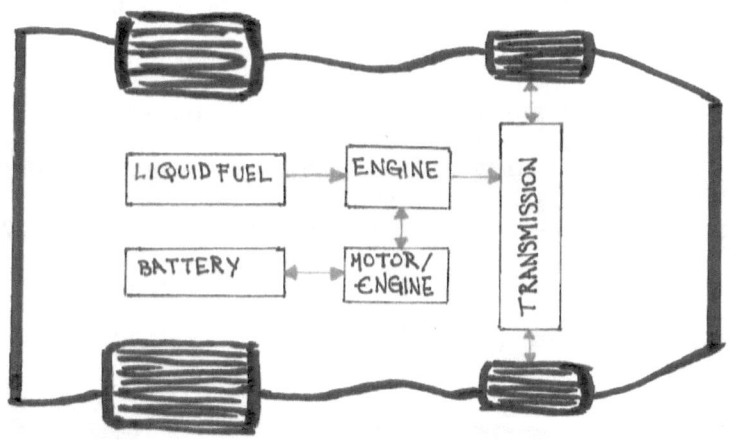

Figure 29 – Mild Hybrid Powertrain

In fact, the advantage of such a configuration is that the motor/generator can be used to switch the engine off instead of keeping it idling while the vehicle is stopped. Turning the engine off when the vehicle is at stop saves a few liters of fuel – 3 to 5% fuel savings depending on the vehicle and the driving cycle or conditions.

Another advantage of such architecture is to assist the combustion engine in situations where additional power is needed and/or let the ICE operate at its sweet spot efficiencies in some specific load points of the driving cycle. The ICE would, in this case, operate at a lower load than what is expected to achieve the driving condition, which burns less fuel and pollutes less, and the electric motor (ISG) will compensate the difference by assisting the ICE and compensate for the load difference. Similarly, at lower loads driving conditions, the energy management system could increase the load on the engine in order to recharge the electric battery. Figure below illustrates well the situation:

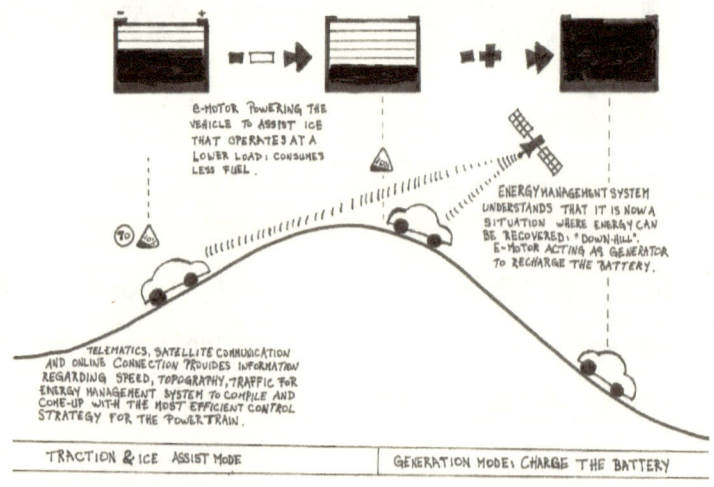

Figure 30 – ICE assist mode in Mild Hybrid Vehicle

SERIES HYBRID ELECTRIC VEHICLES

In a series hybrid there are two energy sources, but a single path to power the wheels of the vehicle. As shown in the illustration below, the fuel tank feeds an engine which is coupled to a generator to charge the battery that then provides electrical energy to a motor/generator to power the wheels through a transmission[25]. The motor/generator is also used to recharge the battery during the sections of the driving cycle when energy can be recovered by deceleration and/or braking.

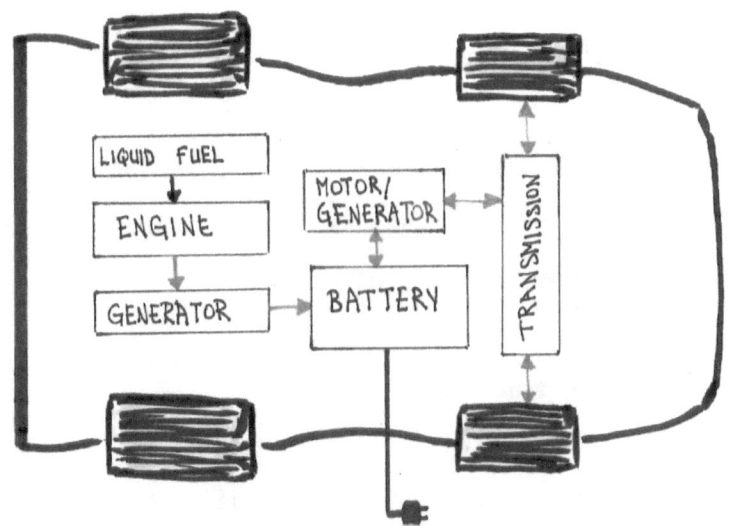

Figure 31 – Series Hybrid Powertrain

---

[25] A direct coupling can also be used. It is the case when we use an In-Wheel or hub motor.

Depending on the components' size, the driving cycles, and charging opportunities from the grid, a series hybrid powertrain can operate in the following several modes:

a) Engine traction only

The vehicle could operate on engine traction only. This means that the engine coupled to the generator would follow the driving cycle load to generate the exact amount of power needed by the traction motor to achieve the duty cycle.

b) Electric traction only

The vehicle could operate in electric traction only. In this case the vehicle works as if it was a pure electric vehicle or BEV. The ICI is off.

c) Hybrid traction

The vehicle could operate in electric traction only. In this case the vehicle works as if it was a pure electric vehicle or BEV. The ICI is off.

d) Engine Traction and Battery Charging

In this mode the gen-set generate more energy than what is drawn from the battery to achieve the driving cycle, so we end-up with an SOC at the end of the driving cycle which is higher than the SOC of the beginning of the driving cycle. For example: Let's suppose that to perform a given driving cycle we need 2kWh of energy. And when the vehicle starts the cycle, the battery of the series hybrid vehicle – which is an 8kWh battery pack is half-charged – 50% SOC (State Of Charge) so 4kWh of energy available. If we assume a 50/50 share between the battery and the gen-set in this hybrid mode, we end up at the end of the cycle with a 3kWh of energy available in the battery. Yes, 2kWh has been drawn from the battery to achieve the driving cycle, but during the cycle the get-set also pushed 1kWh of energy in the battery – half of what was needed to perform the driving cycle under this mode.

e) Battery Charging and No Traction

In this mode the e-motor is off, there is no traction; we just turn the gen-set on to charge the battery.

f) Regenerative Braking

In this mode, we use the electric motor to brake the vehicle instead of using the traditional hydraulic brakes. By reversing the flow of current inside the motor a torque is applied at the opposite direction of traction on the wheels and therefore brakes the vehicle while charging the battery. Kinetic energy of the vehicle is transformed into electric energy in the battery. And we have the double advantage of achieving our driving request – stopping the car – and recovering the energy of the car back for future use.

All these operating modes are as many degrees of freedom to operate the vehicle in the most optimized condition in order to maximize safety, performance, and fuel savings.

Although most series hybrids use an ICE, it is also possible to design a series hybrid using a fuel cell powered by hydrogen, creating a Fuel Cell Electric Vehicle (FCEV). Fuel cell is another alternative to combustion engines for battery charging or so-called range extending. Instead of using fossil fuel like ICE, a fuel cell is using hydrogen as mean of producing electricity to charge the battery.

PARALLEL HYBRID ELECTRIC VEHICLES

In a parallel hybrid vehicle, there are two parallel paths to power the wheels: one fossil fuel path and one electrical path, as shown in the illustration below. The transmission, with multiple inputs couples the motor/generator and the engine, allowing either, or both, to power the wheels.

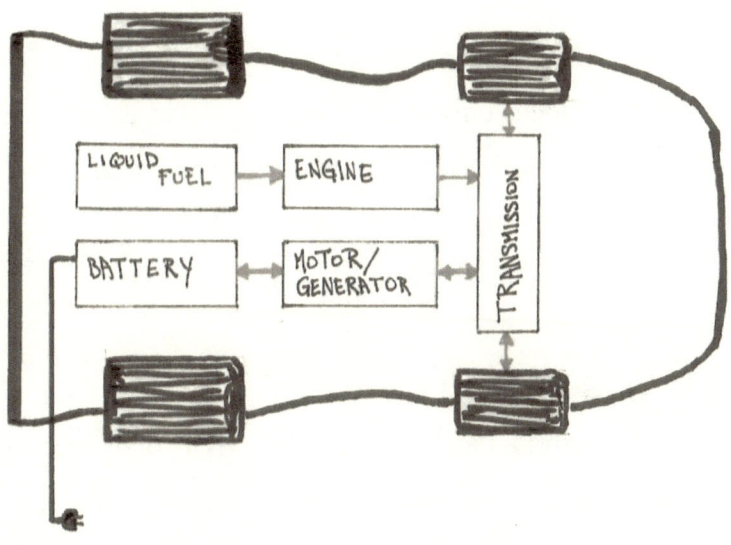

Figure 32 – Parallel Hybrid Powertrain

The advantage of the parallel hybrid architecture is the possibility to operate in many different modes, while maintaining – with a relatively small battery pack –1 to 4 kWh - the cost of the vehicle at an acceptable level[26]. The operating modes with a Parallel Hybrid are as follow:

a) Engine traction only

In this mode the ICE is the sole powering means to the vehicle, this mean that the e-motor/generator needs to be de-coupled from the transmission in a way or another - generally using clutches.

b) Electric traction only

In this mode, only the electric motor is coupled to the wheels through the transmission. This means that the combustion engine has to be decoupled from the transmission through a clutch.

c) Hybrid traction

In the hybrid traction mode, both ICE and e-motor are driving the wheels making it possible to split the load between the two motors. The amount of load each one bears will depend on what is the best split for the better performance and efficiency the system is capable of

---

[26] When a breakeven is achievable between 1-3 years from fuel and maintenance savings it is a no brainer for the consumer. Paying a premium for the EV version is not a barrier anymore.

delivering.

d) Regenerative Braking

In this mode, we use the electric motor to brake the vehicle instead of using the traditional hydraulic brakes. By reversing the flow of current inside the motor a torque is applied at the opposite direction of traction on the wheels and therefore brakes the vehicle while charging the battery.

e) Battery charging from the engine

In this mode the electric motor will be driven by the ICE through the proper clutching of the transmission and will generate electricity back to the battery. In this mode the e-motor is decoupled from the wheels.

Going through this clutching unclutching of the ICE and e-motor to switch from one control mode to the other provides a lot of flexibility to maximize driveability and performance. However, this makes the control of a parallel hybrid much more complex than that of a series hybrid. This complexity has often a big impact on the cost of the vehicle, like the Prius or the Chevy-Volt.

SERIES-PARALLEL HYBRID ELECTRIC VEHICLES

Finally, there is the Series-Parallel HEV. In this kind of architecture, the vehicle has both series and parallel energy paths. As shown in the illustration below, a system of motors and/or generators that sometimes includes a gearing or power split device couples allows the engine to recharge the battery. Variations on this configuration can be very complex or simple, depending on the number of motors/generators and how they are used. These configurations can be classified as Complex Hybrids (such as the Toyota Prius Hybrid), Split-Parallel hybrids, or Power-Split hybrids.

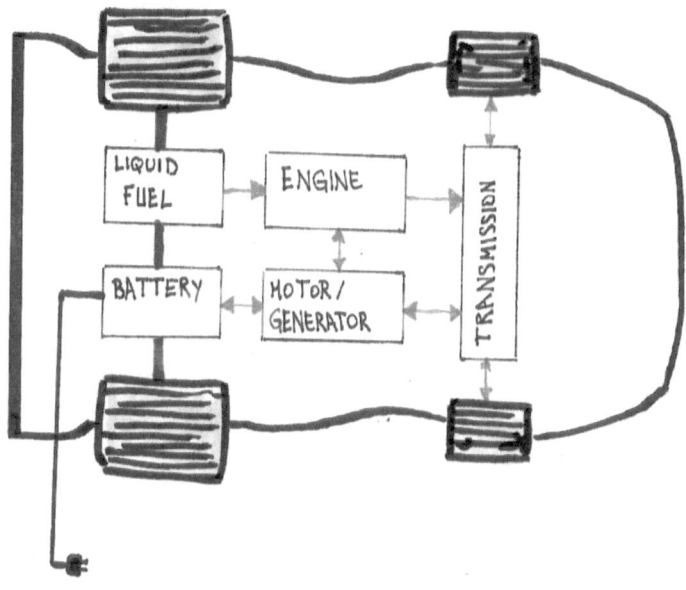

Figure 33 – Series-Parallel Hybrid Powertrain

Another vocabulary that you may come across and which is worth it to introduce is PHEV. PHEV refers to plug-in hybrid electric vehicle. It is an HEV that can be plugged-in or recharged from wall electricity. PHEVs are distinguished by much larger battery packs when compared to other HEVs. The size of the battery defines the vehicle's All Electric Range (AER), which is generally in the range of 30 to 50 miles. PHEVs can be of any hybrid configuration. As you will see in the next chapter, many

PHEVs are available on the market today. Let's have a little tour.

CHAPTER 5

WHAT'S ON THE MARKET

Whenever we mention electric cars, price is generally what pops-up in the mind first, because they are more expensive than their regular counterparts.

The few pages below list each of the plug-in cars on sale today. For each vehicle you will see the manufacturer's suggested retail price, plus any mandatory destination and handling fees but excluding any local or federal tax incentives or rebates. This means that many of these models here may be available cheaper, for those eligible for specific credits or rebates.

A few plug-in vehicles have been left from this list, for example the 2017 Chrysler Pacifica Hybrid and 2017 Chevrolet Bolt EV, which doesn't start production until the end of the year.

The Honda Accord and Toyota Prius plug-in hybrids have also both been removed from this list, as both models were discontinued.

There are several plug-in hybrids that aren't on sale yet while we are writing these lines, but will certainly be in the coming months, including the 2017 Audi Q7 e-tron, 2016 Mercedes-Benz GLE550e, and C350e, and the BMW 330e. Hyundai will also introduce its Ioniq later this year, with hybrid, plug-in hybrid, and battery-electric variants.

The cars are listed from the cheapest to the most expensive and MPGe[27] figures listed below refer to the cars' electric efficiency.

Here is what you can buy today:

---

[27] Miles per gallon gasoline equivalent (MPGe or MPGge) is a measure of the average distance traveled per unit of energy consumed.

- **Mitsubishi i-MiEV - $23,845**

62 miles (EPA), 16 kWh battery, 49 kW motor, 112 MPGe

The i-Miev has never been a strong seller in the U.S, probably because of its look and low performance. But with the recent slash in its pricing make it the cheapest electric vehicle on the market. Some owners could drive from a dealership having paid less than $16,000 for their i-MiEV, if they're able to maximize their use of incentives.

- **Smart Fortwo Electric Drive - $25,750**

68 miles (EPA), 17.6 kWh battery, 55 kW motor, 107 MPGe

The Smart's Fortwo Electric Drive is one of the less expensive new electric cars on the market, but you only get two seats. There's enough power to make driving experience, and if you're able to benefit from incentives, the price starts to look reasonable. Convertible models are an extra $3,000 but top-down electric driving is a wholly pleasant experience.

- **Chevrolet Spark EV - $25,995**

82 miles (EPA), 19 kWh battery, 105 kW motor, 119 MPGe

Only sold in California, Oregon, and Maryland, the Spark EV will soon be replaced by the 2017 Chevy Bolt EV, a 200-mile electric car that will be sold in across the US and Canada. Chevrolet has put the same effort into this model as it did with the Volt, and has managed to improve the aerodynamics and interior to match the expected performance. The Spark EV is good fun to drive.

- **Volkswagen e-Golf - $29,815**

83 miles (EPA), 24.2 kWh battery, 85 kW motor, 116 MPGe

The e-Golf is still only available in a few states, but at least there's now a second trim level that drops the price a bit. The addition of this new SE model amounts to a roughly $4,500 cut in the base price. Range is almost identical to the base Nissan Leaf.

- **Nissan Leaf - $29,860**

84 -107miles (EPA), 24-30 kWh battery, 80 kW motor, 112-114 MPGe

Five years after its debut, the Leaf remains the bestselling electric car in history. Nissan will probably exceed the 250,000's units sold this year, a testament to the car's wide appeal. For 2016, the Leaf gets an optional 30-

kWh battery pack that boosts range to 107 miles.

- **Ford Focus Electric - $29,995**

76 miles (EPA), 23 kWh battery, 107 kW motor, 105 MPGe

Featuring lower range than the Nissan Leaf, but it compensates with more power for those looking for better driving experience. Sales have thus been very slow for this model. Ford is putting more faith in its other plug-in models, the C-MAX and Fusion Energi. Ford says it will add DC fast charging and increase range to 100 miles for 2017[28], so it might be worth waiting for the updated model.

- **Ford C-MAX Energi - $32,645**

20 miles (EPA), 7.6 kWh battery, 88 kW motor (195-hp combined), 88 MPGe

Like the Toyota Prius V, it's a very practical car, ready to handle everything related to a family lifestyle. This car mixes good performance with very impressive efficiency in electric mode. Many users end-up using it in electric mode only 95% of the time.

---

[28] http://www.greencarreports.com/news/1101359_updated-2017-ford-focus-electric-100-mile-range-dc-fast-charging

- **Fiat 500e - $32,780**

84 miles (EPA), 24 kWh battery, 83 kW motor, 112 MPGe

Probably a better vehicle than the gasoline version. Fiat's 500e electric car may be a mere "compliance car", but the engineers have really done a great job. It's fun to drive but the limited availability is an issue. The price is pretty steep for such a small car even if it seems that Fiat does not make any profit on this model!

- **Kia Soul EV - $32,800**

93 miles (EPA), 27 kWh battery, 81 kW motor, 105 MPGe

Like with the e-Golf, the success of the Kia Soul EV probably comes from the fact that it's based on a popular existing model. The Soul EV isn't available nationwide right now, but its relatively long range has created much consumer enthusiasm where it is sold.

- **Chevrolet Volt - $33,995**

53 miles (EPA), 18.4 kWh battery, 111 kW motor, 106 MPGe

The model benefits from a complete redesign for the

2016 model year, including an all-new powertrain, and a battery pack that grows from the previous 16.5 kWh to 18.4 kWh. That allows for an increased electric-range from previous model reaching the 53 miles. This year's model is also more pleasant to look at than its predecessor, with a fifth "seating position," even if not one that's useful for long trips.

- **Ford Fusion Energi - $34,775**

20 miles (EPA), 7.6 kWh battery, 88 kW motor (195-hp combined), 88 MPGe

With similar architecture to the C-Max, this means the same battery electric range and efficiency rating, despite the two different body styles. The Fusion is the best looking of the pair though. But all that extra metal means finding some more extra cash before you sign on the line. More extensive upgrades are planned for 2017.

- **Hyundai Sonata Plug-In Hybrid - $35,435**

27 miles (EPA), 9.8 kWh battery, 50 kW motor (202 hp combined), 99 MPGe

This is the Hyundai's first-ever plug-in hybrid version of the mid-size sedan. That adds new rivals to the Ford Fusion Hybrid and Fusion Energi twins. A plug-in hybrid version of the related Kia Optima is also expected soon as well to add more choice to this segment.

- **Audi A3 Sportback e-tron - $37,900**

16 miles (EPA), 8.8 kWh battery, 75 kW motor (204 hp combined), 83 MPGe

First production of Audi plug-in hybrid sold in the U.S. It's also the only way to get an A3 hatchback on this side of the Atlantic. Its restrained and very casual styling is perfect for those who don't want to attract too much attention to them.

- **Mercedes-Benz B250e - $42,375**

87 miles (EPA), 28 kWh battery, 132 kW motor, 84 MPGe

This Mercedes hatchback is only available in certain electric-car friendly states, but serves as a more practical alternative to the BMW i3 at a pretty much similar price.

- **BMW i3 - $43,395**

81 miles (EPA - i3 REx 72-150 miles), 22 kWh battery, 125 kW motor, 124 MPGe (i3 REx 117 MPGe)

Range-extended models begin at $47,245. The i3 is the most energy-efficient vehicle sold in the U.S. Even if it's one of the more expensive BEVs on the market aside from the Tesla, it's also one of the best. Its carbon fiber-reinforced plastic construction is very different from what you could find in the segment, so is its futuristic styling and loft apartment-style interior.

- **BMW X5 xDrive 40e - $63,095**

14 miles (EPA), 9 kWh battery, 82 kW motor (308 hp combined), 56 MPGe

Fifth plug-in vehicle from BMW, this plug-in hybrid SUV competes with luxury plug-in hybrid SUVs from Porsche and Volvo, and will soon be joined by competitors from Audi and Mercedes-Benz.

- **Cadillac ELR - $64,995**

40 miles (EPA), 17.1 kWh battery, 174 kW motor, 85 MPGe

Based on the first-generation Chevy Volt, the ELR has not been a commercial success at launch. So Cadillac decided to cut its extremely high base price, boosted power, and added upgraded suspension and other new features.

- **Volvo XC90 T8 "Twin Engine" - $69,095**

14 miles (EPA), 9.2 kWh battery, 60 kW motor (315 hp

combined), 53 MPGe

The XC90 uses a turbocharged and supercharged four-cylinder engine from Volvo's "Drive-E" line, paired with electrification for the first time. It is the only seven-seat plug-in hybrid available, and one of an emerging breed of luxury plug-in SUVs.

- **Tesla Model S - $71,200-$106,200**

234-270 miles (EPA), 70-90 kWh battery, 284-568 kW motor 89-101, MPGe

The Tesla Model S recently updated the lineup with its "Autopilot" and "Summon" autonomous driving features, and a "Ludicrous" mode for performance models. One should be very careful using these feature, especially in light of the recent accident that caused the death of a Tesla model S autopilot user.

There's also a 90-kWh battery pack option that Tesla says should increase range by 6 to 7 percent, but that doesn't seem to show up in EPA testing.

- **Porsche Cayenne S E-Hybrid - $78,250**

14 miles (EPA), 10.8 kWh battery, 70 kW motor (416 hp combined), 47 MPGe

For the 2015 model year, Porsche is offering a plug-in

hybrid powertrain in its Cayenne SUV. This will give Porsche a head start over the upcoming plug-in SUV competition.

- **Tesla Model X - $81,200**

220-257 miles (EPA), 70-90-kWh battery, 568 kW motor (P90D), 89-92 MPG,

Looking for a seven-seat all-electric vehicle that can tow a large caravan? Your choices then are pretty much limited to the Tesla Model X. With its attention-grabbing "Falcon style" and powertrains adapted from the Model S, the Model X is an impressive first attempt at a crossover from Tesla.

- **Porsche Panamera S E-Hybrid - $94,250**

16 miles (EPA), 9.4 kWh battery, 70 kW motor (416 hp combined), 51 MPGe

The Panamera S E-Hybrid is really more of a sporty and fast luxury car with supplemental electric power than a truly green vehicle. Like a Tesla Model S, Porsche's first mainstream plug-in hybrid features four doors and the expected sporty attitude.

- **Mercedes-Benz S550e - $96,57**

14 miles (EPA), 8.7 kWh battery, 85 kW motor (329 hp combined), 58 MPGe

The Mercedes-Benz S550e is a plug-in hybrid version of the company's big luxury sedan. It is definitely offering the best of both worlds, but in reality the S550e is likely more focused on luxury and performance than outright fuel efficiency. Like for all cars in this range of price, decision making process isn't always straight forward for the regular Jacks and Pauls!

- **BMW i8 - $141,695**

15 miles (EPA), 7.1 kWh battery, 96 kW motor (357 hp combined), 76 MPGe

Very different from the i3, the BMW i8 is a smooth plug-in hybrid coupe with the similar carbon fiber-reinforced plastic and aluminum construction to the i3. The hot styling is certainly backed by the emphasis on performance. Why would you pay that much for a car, if it could not be both good looking, and fast and efficient?

PART 3 – ENTER THE BOOM!

MOHAMMED BELBARAKA

ENTER THE BOOM!

Despite the advances that vehicle electrification has made in the past few years, there are still significant barriers that stand in the way of a widespread adoption of EVs. Technological, financial, market, and policy challenges could hinder mass market penetration of EVs if all stakeholders do not work in harmony. More collaboration is needed for further R&D investments and demonstration programs. More public-private partnerships and innovative policy will also help the development and expansion of EV markets.

In this section we will try to identify the most important challenges to high volume deployment and highlight the opportunities that governments, in coordination with the private sector and the broader EV stakeholder community, can pursue to make the positive impact expected from mass market penetration of EVs. In fact, many EV challenges can be locally-specific, but those exposed below are some of the major issues facing both early market leaders and countries still contemplating initial approaches to electrification.

Before we dive into the development of the major challenges still on the way, it would be worth it to stop and have a look at what has been achieved so far. What is the current status of the EV industry globally?

CHAPTER 6

INVENTORY OF FIXTURES

Last year, according to the International Energy Agency (IEA) the global EV stock has for the first time in history hit the threshold of 1 million vehicles on the road, closing at 1.26 million[29] according to the same source. We are still far from the 20 million vehicle objective by 2020 but still, this is a symbolic achievement highlighting significant efforts deployed jointly by governments and the industry players over the past ten years. Only two years ago about

---

[29] This figure and the objective of 20 million is for plug-in passenger vehicles. It includes all PHEVs, BEVs and FCEVs.

half of today's EV stock was on the roads, and farther back in 2005, electric cars were still measured in hundreds. It is indeed a huge step toward for more significant EV market introduction.

Between 2008 and 2014, the 16 billion USD[30] investments in R&D, charging infrastructure and fiscal incentives' programs definitely helped the industry development and progress. In many countries, buying an EV is becoming a more and more easy experience within reasonable reach to the general public. The market shares of EVs is quite significant in counties like Norway, France, Sweden, Denmark, UK and China who's vehicle electrification is well diversified - from electric two-wheelers to electric buses.

In parallel with the growth of the global EV stock, a substantial improvement of the charging infrastructure has been observed. In fact according to the Electric Vehicle Electrification Initiative[31] (EVI) group's publications EVSE (Electric Vehicle Supply Equipment) - in other words charging spots - have more than doubled for slow charging points since 2012, and multiplied by eight for fast

---

[30] Source : IEA –Global EV outlook 2015
[31] The Electric Vehicles Initiative (EVI) provides a forum for global cooperation on the development and deployment of electric vehicles (EVs).

charging in the same period. Public policies are encouraging publicly accessible charging development through direct investment and public-private partnerships. Such partnerships are occurring in urban areas and beyond, with charging networks aiming to enable long-distance travel on EVs even at the continental scale, as in the case of Europe and North America.

Industry, governments and early adopters have played an important role in demonstrating that electric cars can deliver the practicality, sustainability, safety, and for some EV models now available, the affordability characteristics expected from them. Nevertheless, the EV market still requires strong policy support to achieve widespread adoption and deployment. Battery costs have been cut by four since 2008 and are on the right path to decrease even further. At the same time, battery energy density still needs to increase to enable longer ranges at lower prices. (This will be further detailed in the next chapter.)

Technological progress, public awareness and economies of scale are critical to move towards cost parity with conventional ICE propelled vehicles. Recent automaker announcements suggesting EV ranges that will soon be exceeding 300 kilometres (km) give encouraging signals for the future.

Other transport modes are getting more and more electrified. Transit buses, and all sort of delivery vehicles, or even 2-wheelers, are currently developing particularly in China where the estimated stock of all EVs is exceeding 200 million units.

With a strong commitment to electrification at both national and local levels which translates into strong fiscal and infrastructural policies, China has emerged as the global leader in the electric 2-wheelers market and is almost the only relevant player globally today. Policies like the restriction on the use of conventional two- wheelers in several Chinese cities have both helped reduce local pollution and noise, and at the same time propelled this new EV industry in the country. Deployment of electric bus fleets is another example of the successful Chinese policies that pushed the country to the rank of the global leader in this segment with more than 170 000 buses already circulating today. China is considered an Eldorado for many western companies that are chasing opportunities in the country to sell their products.

Most Essential

- **The electric car stock has been growing since 2008, with the BEVs ahead of PHEVs.**
- **80% of the electric cars on road worldwide are located in the United States, China, Japan, the Netherlands and Norway.**
- **Current development in battery technologies gives good confidence to improve both energy density and costs according to automakers and the US DOE.**
- **Meeting the very ambitious decarbonisation and sustainability goals requires a major deployment of electric vehicles – 20 million+ - from now to 2020**

With only 1.2 million electric vehicle on road today, there is certainly a lot of room left for the boom to happen! Let's talk about this ocean of new opportunities.

CHAPTER 7

PREPARING FOR TOMORROW

TECHNOLOGICAL AND POLICY CHALLENGES

Battery limitations

As mentioned earlier in the general electric vehicle architecture discussion, the most significant technological challenges currently facing EVs are the cost and performance of their components, particularly the battery. The price per usable kilowatt hour of a lithium-ion battery, depending on volumes and Original Equipment

Manufacturer's (OEM) supply chain strategies, ranges between $300-$600 USD and thus represent a significant portion of a vehicle's cost, specifically BEVs which have large battery packs. A Nissan LEAF, for example, has a 24 kWh battery pack and costs approximately $30,000 USD with about a third of the vehicle's retail price covering only the battery pack. With 6kWh less than the Leaf, Chevrolet Spark sells for $26,000, here again a significant part of the price goes to the battery pack. Some PHEV models are even more expensive due to the cost and complexity of their dual powertrains. A Chevrolet Volt for instance uses only an 18 kWh battery pack, but its purchase price is nearly USD 4,000 more than a LEAF, due in large part to its hybrid technology. On the high end of most fancy EV offerings, Tesla is the THE example of success stories from design, performance and even retail price per kWh of embedded energy perspectives. In fact, if you compare the ratio of retail price to kilowatt hour of battery pack, the Tesla Model S is less expensive than a Nissan leaf - 1160$/kWh for the Tesla Model S vs 1245$/kWh for the Nissan Leaf.

Since the introduction of the first EV models to now, prices remain more expensive than their petrol vehicle equivalents even when combined with government purchase subsidies offered in many countries. In fact, according to the IEA (International Energy Agency), twelve EVI (Electric Vehicle Initiative) member governments offer some type of fiscal incentive at the national level for purchasing electric vehicles, usually in the form of tax credits or direct rebates. Many governments cap purchase subsidies at a certain amount of money or manufacturer sales volume. Unfortunately, some are scheduled to expire soon. As government subsidies begin to phase out, the upfront purchase price will be at very high levels unless substantial cost reductions are achieved either with high volumes or with the introduction of a breakthrough battery technology.

In fact, progress through R&D has led to a steadily decreasing cost of battery pack these past years as a result of both public and private sector advances and will likely drop even further in the next coming years due to pack design optimisation and cell count reduction, lower cost of cell materials, economies of scale, and improved manufacturing processes.

According to a multitude of sources, the cost of powering an electric vehicle must be less than 150 per kilowatt hour in order to compete with current conventional vehicles running on diesel or gasoline. A recent publication of news and information website strartfor.com, claims that Nissan and Tesla's cost is close to $300 per kilowatt hour today. This very low price compared to what other small players have to pay today for a battery pack – between $800 and $1200 per kilowatt hour – is probably due to their global supply chain strategy and full control over a larger portion of the battery packs chain of value. Economies of scale and removal of middle-men helps indeed reducing the cost. In their global EV outlook 2016 publication report, the EVI group confirms the values above, but adds an interesting information regarding the evolution of energy density. In fact, battery technology has progressed constantly in terms of safety, power, cost but also energy density – from 50Wh/liter in 2008 to 300Wh/liter now. This means that with the same battery size, the range increases 6 fold or the other way around, the same range could be achieved with a battery pack six times smaller than the one of 2008.

You may be a little bit confused here. Why the "pack" after battery? A battery is a battery, I have one in every single of my personal devices from smart phone to laptop?

What you need to understand here is how a battery "pack" of an electric car is made and what actually makes it so challenging to build it, to the point that it is standing pretty much as the number one roadblock to mass market electrification.

It is challenging because of the very nature of the battery pack and its behaviour- both when it is being charged and when it is discharged. A battery pack is made out of modules assembled together to constitute the final battery pack featuring the energy capacity requested for the vehicle it is built for. Each module is made out of single cells bundled and packaged together to form the module. At this stage, the hurdle starts. When battery cells are bundled together - in order to multiply the power and the energy capacity of the single cell- the behaviour of each cell is different from the other group of cells. If proper control and monitoring is not implemented at the cell level, the module will be prematurely damaged. At this stage, along with cells, additional electronic components are added like voltage and temperature sensors in order to monitor each of the cells' condition and be able to control the charge and discharge rates. This level of complexity is further accentuated at the pack level when several modules are bundled, thus another layer of electronic components, control boards and complex control algorithms to manage the energy flow in and out of the pack. Each battery pack comes with its Battery Management System (BMS) which is basically an Electronic Control Unit (ECU) that monitors the energy flows according to both the traction system's requested -energy- and the physical status of each module in the pack. Depending on the internal condition of the pack, the BMS might impose a current limitation in order to preserve the integrity of the battery pack in some specific conditions.

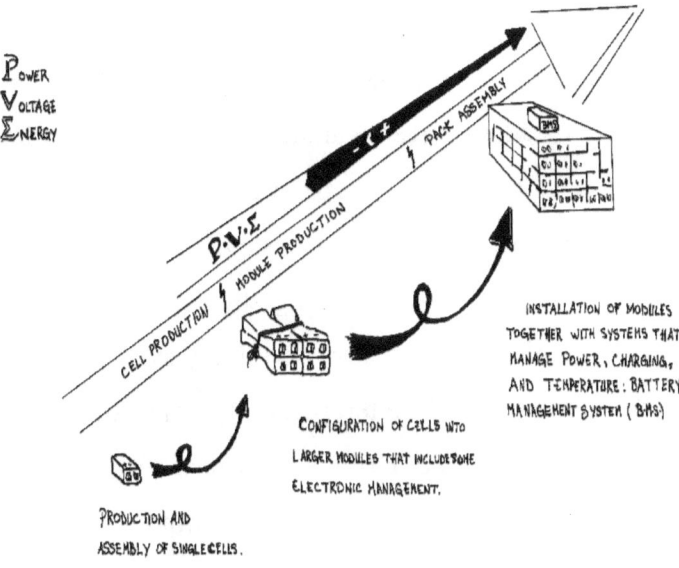

Figure 34 - Value Chain of Electric Car Battery Packs

Charging infrastructure limitations

The next significant challenge to mass market introduction of EVs is the availability of a well suited charging infrastructure. The lack of harmonised standards and interoperability between different charging systems makes it difficult for EV users to find the appropriate charging location for their vehicles. It becomes crucial to develop a common set of standards for charging couplers and communications protocols to help market development. This will lead to lower manufacturing costs and provide seamless and predictable operation for EVs to the end-users. Interoperable charging systems will allow any EV user with any EV model to charge their car at any charging spot, regardless of operator or billing system. Currently standardisation and interoperability are of particular importance in Europe. Drivers must be sure they can drive from one city or country to another without encountering incompatible charging networks before we could see a significant increase in EV sales.

On a global scale, fast charging systems currently face competing standards, one being the CHAdeMO protocol adopted by Japanese industry and the other being SAE International's Combined Charging System (CCS) adopted by U.S. and German car manufacturers. Fast charging installations are more powerful and use different technologies than the regular wall supplies. The advantage is that it could charge the battery at a higher power rate, thus in shorter times. It is indeed more complex and more expensive infrastructure.

So, in order to avoid a costly proliferation of parts and software, consistent standards should be developed through established standards development organizations such as the International Electrotechnical Commission[32] (IEC) and the International Organization for Standardisation[33] (ISO). Other market-driven solutions may be needed to achieve as much compatibility as possible between existing standards. Government support for industry-led voluntary standards efforts is important, as is international collaboration on EV standards harmonisation. Many countries have already started cooperation on standards through multilateral and bilateral EV initiatives[34].

---

[32] (IEC) is the world's leading organization that prepares and publishes International Standards for all electrical, electronic and related technologies.

[33] ISO is the International Organization for Standardization. It develops and publishes International Standards.

[34] GreenCarReports.com, 19 November 2009

Range and safety concerns

The sizable EV price premium would perhaps be acceptable to a large number of consumers if the vehicles offered more range or differentiated functionality than is currently on the market. With a usable range of about 100 to 150 kilometres (km), the 19 kWh battery-powered Chevrolet Spark EV achieves about a fifth of the range of a comparable ICE vehicle. All-electric vehicles with larger battery packs, such as the 85 kWh Tesla Model S offer much greater range (480 km) but also come with a significantly higher retail prices, which not all consumers can afford. PHEVs eliminate range constraints, but many only offer about 15-65 km of electric-only range and thus may not fully deliver the benefits of an electric drive- such as cheaper fuel and lower emissions- if driven predominantly in fossil fuel-mode.

These range limitations along with the premium cost, even with the considerable technological progress on batteries since 2010, are still holding back many potential customers. Even now in 2016 you can still read comments and posts in many of the EV forums from people who still consider range to be a major disadvantage of EVs. A 2009 Survey from U.S. Dept. of Transportation, National Household

Transportation states that in the United States, the average daily vehicle distance travelled per person is 46 km and average vehicle trip distance is 15 km. Given the fact that U.S. average travel distances are among the longest in the world, it is likely that most of today's electric vehicles have sufficient range for the majority of consumers worldwide. Nonetheless, as long as this gap remains between range expectations and actual average driving needs, negative perceptions about EV range and notions of range anxiety will persist. This highlights one important lever that stakeholders from different sides need to address: public education.

Public education needs to cover all areas of misconception and fear. Safety and reliability are one of the areas where stakeholders need to communicate more. In fact, perceptions regarding the safety and reliability of EVs could also block the way to a bigger market. Fire-related incidents in China and the United States back in 2011, along with other reports of battery failures, recalls, and climate-related battery degradation the following year, has attracted high-profile media attention and further raised doubts about EV technology[35]. More recently, the sad

---

[35] Bradley Berman, "Tesla Battery Failures Make 'Bricking' a Buzzword," New York Times, 2 March 2012; "A123 Systems to post $125 million first-quarter loss after recall," Reuters, May 11,

accident of a Tesla Model S driver running on "auto-pilot" mode raises even more the concern about these new energy vehicles' safety. It is difficult to have an accurate estimation, but one could argue that this has, to some extent, prevented a significant part of the EV sales so far. Again, there is a general consensus that public understanding of electric vehicles is one of the major pillars towards more transport electrification. However, in many developed markets, promoting greater understanding of the benefits and potential of EVs is not simply about giving people more information, but rather a need to challenge commonly held misconceptions, scepticism, and bias. The first and most basic condition is that before an individual can consider buying an EV, they first need to understand that it is a valuable option. The next challenge is to explain the relative benefits of going electric and how these new types of vehicles could match their different needs and lifestyles.

World's first fully-electric racing series, sponsored by the Fédération Internationale de l'Automobile (FIA) – much more known as FORMULA E- is an initiative with the potential to both challenge negative stereotypes and to also

2012. Nikki Gordon-Bloomfield, "Nissan Buys Back LEAFs Under Arizona Lemon Law," Christian Science Monitor, 30 September 2012.

generate positive mainstream news headlines. An informed public which is positively disposed to EVs requires an active education and promotion. While motorsport alone will not challenge all public misconceptions about EVs, it is still a significant contribution to a wider change in the way that EV communications are delivered.

POLITICAL CHALLENGES

Many policies initiatives to help transport electrification have so far been implemented but there is still a lack of clear regulation on some areas that is contributing to slowing down the mass market introduction of EVs.

Some of the policy opportunities to jumpstart the market that can be pursued by governments today are listed below:

Dedicated signage for EVs

Signage for charging infrastructure is another area worthwhile to look at very closely and standardize. Standard signage for EVs is still lacking in many countries and when existing, we see very little consistency across different countries especially in Europe. As irrelevant this issue might seem, it is causing a lot of confusion among drivers and is definitely holding back many undecided consumers from purchasing an EV.

Figure 35 - Example of dedicated signage for EVs

A consistent and abundant signage will do both, enhance driving experience of current EV drivers and encourage potential EV buyers to buy their first EV.

EV signage is a strong means of communication and promotion for EVs at a relatively modest cost.

Availability of charging spots for apartment buildings

A large share of potential EV users in dense urban areas is still out of reach. While the convenience of home charging is one of the most attractive attributes of EVs, such convenience does not easily apply to drivers living in apartment buildings or other multi-unit dwellings (MUDs) that either lack garages or do not have the ability to install charging spots easily. Here is another opportunity for local governments to help increase EV users, and thus improve the air quality and noise levels in high dense urban areas. Public authorities must extend subsidies to MUDs or amend building codes and laws to make EVSE capability mandatory for all new constructions.

Implement stronger fuel economy regulations

Implementing stronger fuel economy regulations provides car manufacturers with incentives to invest in EV technology, among other fuel-efficient technologies, and helps increase product diversity. Furthermore, providing for transparent and predictable fuel economy regulations in the future will help manufacturers prepare to meet them. Governments at the national, regional, and local levels can more directly spur sales of EVs through large-scale fleet procurement. Such high-volume purchases can accelerate economies of scale, while allowing governments to lead by example and perhaps inspire other fleet operators to consider electrification.

Rethink the urbanisation

Mass market introduction will not happen without the optimisation of urban structures to reduce trip distances and shift mobility towards public transportation which of course need to be electric. The EV deployment should happen across all modes of transport in an urban infrastructure.

This is where we realize that the electric boom is extending way beyond the regular spheres of the old conventional transportation sector. It is actually a revolution that is forcing us to reconsider every aspect of our daily lives which will eventually lead to a complete change in current paradigms.

What about the electricity source?

Did you know that with a battery pack of a Tesla Model S you have enough energy to power a family house of five people for 24 hours?

With significantly large amount of energy on-board, an electric vehicle, when connected to the grid, can become a flexible and on-demand asset to enable more reliable and efficient running of electricity systems. Since electricity is bi-directional and with the new improvements in bi-directional chargers, the battery of an EV can now be used as a power supply. Yes, you could use the energy to move the car, but when the car is parked and plugged to the wall, instead of taking energy from the grid you can do the opposite: pull it out of the battery and feed-in the grid.

Achieving transport electrification is only half of the equation resolved. The other portion of the equation is how we could reduce dependence on fossil fuels for electricity production. Many countries are promoting the electrification of transport to reduce expensive fossil fuel imports and to satisfy a greater share of their domestic energy needs by exploiting their own natural resources. Reducing road transport's dependence on fossil fuels –

which is by the way finite resource - is part of a global energy security movement which is focused on ensuring long-term access to an uninterrupted and affordable supply of energy. This is where renewable energy, such as wind and solar, comes onto the stage.

Electricity is a form of energy which has to be dissipated and consumed as it is produced. The different kinds of electric loads plugged in the grid – be it household or industrial equipment- are always powered on from electricity which is actually being produced at the very moment of use. This is what makes load prediction on the grid one of the main critical tasks of electricity producers. The production needs to be anticipated at both long term and also to the second. When using fossil fuel, it is easy to match production with anticipated and actual electricity consumption on the grid: the grid operator just burns more or less fuel to follow the load demand.

It is very important that you get a feel for these notions about electricity in order to understand what happens when we deal with renewable sources of energy. The idea to retain from all the statements above is:

> **Electricity has to be consumed by an electrical device or equipment as it is produced. The only electrical device that could return that electricity back is the battery.**

When dealing with renewables, we are not always able to match production with consumption. Since consumption of electricity needs to happen at the very same time as of production, a wind turbine or solar panels needs to be switched off when there is no "consumer" in the grid they are feeding. Even if there is a lot of wind and sun available, if no electrical device can absorb dissipate or consume that potential electric energy available naturally, the only solution for the operator is to switch the production of the renewable field off.

Renewable sources will only be used when an electric load can be plugged, and this is where operational grid constraints force generators to accept less renewable energy than is available.

Transitioning transportation from fossil fuel energy to electricity can create a controllable and flexible demand for electricity that will play an important role in overcoming curtailment of renewable energy. Why? Thanks to the large number of batteries now available with EVs, we can store all this energy from renewables which had no place to go to before. A battery is the only device that can absorb electricity and regenerate it back in the same form. This electricity storage capacity of the batteries offers a perfect opportunity to refuel with green electricity at times that match the intermittent supply from wind, solar or other renewable sources, and to displace highly polluting fossil fuels. With the growing EV stock, even after batteries have reached the end of their useful life, it could still be used as a valuable resource for managing energy grids once stripped-off the aging cars.

This opens the door to a multitude of new applications and business opportunities including: emergency supply during power shortages or shutdowns, replacing diesel generators that power events, leisure activities or remote buildings where other forms of power are absent, helping grid operators to balance demand and supply fluctuations, and offsetting peak building loads to reduce the energy bills of households and business that are charged tariffs based on maximum usage. By bidding into these new markets, EV owners have access to a new revenue source and an opportunity to reduce the total cost of ownership of their investment. These new business opportunities and models were previously unimaginable with conventional cars.

The climate change-related benefits expected from mass introduction of EVs on the roads instead of their conventional counterparts cannot be fulfilled unless they could be charged from a decarbonised grid. This is an additional challenge for countries that are largely dependent on fossil fuels for power generation. EV adoption would still bring immediate benefits such as air quality improvements and reduced noise in highly dense urban areas. But, basically, if nothing is done for a clean grid, what would happen in this case is that the pollution would be shifted out of the cities to concentrate on the power generation areas.

But this could also be addressed with strategic policies of investment to support the transition by increasing the opportunities available to integrate variable renewable energy production capabilities everywhere in the country.

MOHAMMED BELBARAKA

THE LAST CHAPTER

FINAL THOUGHTS

The electrification of the global vehicle fleet is undoubtedly a long-term ambition. With a few exceptions, EV market shares are still below 1% in most major markets due to the high upfront costs, real and perceived range limitations, and the lack of public education. At the same time, there has been considerable progress in the global market, which suggests a positive outlook.

Besides the obvious potential for creating a more sustainable future if we raise the number of EV on roads worldwide. It's also becoming very easy to imagine the future with electric vehicles and get prepared to seize all the opportunities ahead of us.

There are still of course several actions to take before increasing the world EV stock to the 20 million electric vehicles on the road by 2020 set by the EVI group. Stakeholders and those who will be aware of all the opportunities on the way will play different roles. There is and will be even more all along the way of more and more transport electrification, actions to take for everybody and in every country. There is still so much value to create and so many problems to solve that no one individual, country or sector can be left without playing a significant role. Billions of dollars are going to be made.

Joining, following and participating in groups and associations whether at the local, national or international levels like the Electric Vehicles Initiative (EVI) will help facilitate coordination and communication among stakeholders to address the most significant challenges of vehicle electrification. The challenges that is likely to push in the right direction, and bring real added-value and solutions. Opportunities are created everyday, the key to succeed in this industry is to understand the markets and taking the right actions now.

These actions need to be directed towards three avenues to articulate the necessity of bringing the diverse community of stakeholders' efforts together.

The first avenue is of course technology. As described in the previous chapters, there are still many technological challenges to address before we could see a significant shift in electric transport landscape. Batteries need to carry more energy, weight less and cost less. This area in itself represents millions of dollars' worth of market value and thousands if not millions of jobs, both direct and indirect. The other technological challenge relies in building a more structured and integrated electricity infrastructure - both power generation and supply and distribution. More efforts need to be deployed in improving the renewable power generation technologies to make them more powerful, efficient, reliable, and affordable. The same but at a lesser extent is the quality, performance and cost of all other components of an electric car like electric motors, power modules, electronic and communication modules.

The second avenue is policy. It is essential to come up with a deliberate system of principles and guidelines to orient decisions and achieve the expected rational outcomes to the EV's large scale introduction. Technologies' introduction to the public sphere is always challenging the way a society is organized and very often makes the sets of rules and organizations resulting from previous technologies no applicable and obsolete, if not dangerous at all. The electrification of transport should be considered under the context of an increasing urbanisation and population density. Today, more than half of the world's population lives in cities and the UN (United Nations) project the proportion to reach 70% by 2050[36]. For a better quality of life in large cities, a grand scale mobility strategy is mandatory. Improved and expanded public transit, enhanced pedestrian and bicycle access, and new "mobility services" should be the components of such a strategy and policies' implementation. EVs have a major role to play in these green and smart cities, and their technologies have huge potential spillover benefits for a variety of industries. With strong political dynamics, these effects could be both immediate and long-lasting, thus altering the world for the better.

The technological progress is happening at a very fast pace and it is very often difficult for politicians at all levels to follow or to even understand what is happening, and to what extent the technology could be disruptive. Here we notice the importance of good communication channels and coordination with the other avenues' stakeholders. Policy implementation should happen at different levels from both

---

[36] "U.N.: By '09, Half the World Will Live in Cities," USA Today, 26 February 2008.

the top down and the bottom-up. From budget allocation to finance R&D programs, to infrastructure investment, planning and regulation, all the way to large public communication campaigns and incentive programs. In this area there is also lots of room for the "innovation genius" of politicians and public agents of all kinds and from all backgrounds.

Finally, the third avenue is finance. It will be important to build the appropriate financial tools to support this new growing market. This new industry is coming up with new ways of creating value. The distribution of all these new added values will need new specific and adequate set of financial tools in order for each stakeholder to agree playing within a fair game.

For example, both public and private players could join their efforts to seed charging infrastructure investments in new markets. They could build proposals and strategies to fairly share the heavy upfront deployment costs and anticipated revenues. They could also identify all the possible revenue models like traditional and battery-only leasing, development and regulation of the EV resale market, and develop adequate regulation on utility status of the charging infrastructure service providers. A great example of an innovative financing model could be the development of financial products that combines the advantages of an electric vehicle with the cost savings that can be achieved from a home renewable energy installation like a wind turbine or solar panels. For example, a typical household that will spend on average a total of $1000 per month for home electricity, natural gas, as well as vehicle fuel costs and loan payments, could easily match or undercut those costs with a combination of an electric vehicle, a home energy efficiency retrofit, and a rooftop solar system. This kind of model provides a means to combine the financial benefits of these technologies and services, creating a financial package that rolls all technologies and services into one loan. This loan could then be paid off in less than five years if the banks agree on reasonable interest rates and loan conditions. After those five years, customers never have to pay for petrol for their car or electricity for their house, as long as they live there. This is a new business model with a real value proposition to both drivers/homeowners and financial institutions.

A number of vehicle manufacturers and energy companies around the world have formed alliances to capitalize on the potential of combined products related to electric vehicles. Outside of the energy sector, other innovative business models are also being developed to enhance the EV driving experience. This includes wider personal mobility packages that give EV drivers access to the most appropriate vehicles for different trips, such as vans, bicycles, and petrol vehicles for longer journeys. Electric vehicles and EV charging also offers considerable potential to be bundled with other transport oriented services such as parking, valet, repair, maintenance, navigation, car-sharing, and public transport. Such convenient and cost effective packaging of multiple products could introduce the electric vehicle to new customer segments in both the private and commercial fleet markets, enhancing the benefits and ease of switching to EVs.

These listed examples above are merely an attempt to start the discussion and maybe peek the interest of financial experts who are most certainly needed and have a major role to play.

Despite a long history of repeated ups and downs, electric mobility continues to advance toward a better state of art and a more durable market presence. Indeed, EVs continue to open up a variety of consumer segments not considered possible in the past, thanks to the leveraging of the three avenues mentioned above. The road ahead to meet countries' ambitious sustainability goals will not be easy but will definitely be amazing for those who will have the chance to participate to this revolution. Significant market penetration will unfold gradually over next 10-20 years, thus requiring a healthy dose of patience for those anticipating a new era of clean transport. Transforming the way automobiles are powered, cleaning and scaling up the requisite infrastructure will not occur in a matter of months. The challenges facing vehicle electrification are serious, complex and will therefore necessitate a broad and coordinated effort among all relevant stakeholders to address them. The good news is that we are the ones who will be creating and building this new world full of promises and opportunities, because this is indeed all happening now.

The relative simplicity of electric vehicles' architecture creates opportunities for new players to challenge the long-established dominance of old automotive manufacturers. These companies have previously competed on their ability to engineer internal combustion engines that is one of the few components that are very complex to develop and manufacture. This has provided them a dominant status and created very high barriers to entry to market over small challengers. Now, as electric vehicles only use basic motors and gearboxes that are considerably easier and cheaper to both develop and assemble, more room is left for new players to come-up with innovative proposals.

New players offering niche products have begun to make a mark in the burgeoning electric vehicle industry. Tesla Motors only began producing cars in 2008 and by 2013, its Model S was one of the best-selling electric cars in the United States, with the company significantly outperforming established vehicle manufacturers in the stock market. Another example is BYD, a Chinese automotive manufacturer, which is taking steps to expand outside of China. However, there have also been a number of notable EV start-up failures—such as Fisker, Think, CODA, and Modec—which demonstrate the difficulty of competing in the automotive sector. Nevertheless, the ability of new players to introduce innovative product and service offerings to reduce costs and increase acceptance of electric vehicles could prove an important element in the widespread adoption of this new technology.

Further competitive challenges to old automakers, will also come from the new patterns of mobility that are changing the models of car ownership, and the expected decline in aftersales revenues owing to the relative mechanical simplicity of electric vehicles.

Introduction of EVs is merely a small piece of a large scale revolution unlike mankind has ever experienced. As we have just mentioned above and across the last chapters, the stakeholder's community is more and more extending beyond the mere industry players and is reaching out to areas we would not thought of in the first place.

In fact, the technology is progressing so fast and is generating new ways for us to interact with our cars and our conception of mobility. With the emergence of other technologies like the Internet of Things (IoT), our cars, buses, taxis... we will become more and more connected to the outside world. This connectivity combined with the introduction of driverless cars will open an infinite range of possibilities and opportunities.

Access to infinite amounts of information with more and more sophisticated computers and complex algorithms combined with the possibility to move physical objects across a relatively large perimeter will definitely accelerate the revolution and change of paradigms. Let's highlight this point with some concrete real life examples.

Close your eyes and imagine a world with no more wasted time looking for a free parking spot. Before even making your trip and with devices communicating together effectively, a spot could be programmed or reprogrammed for you even before starting your journey.

No more bothering with the maintenance of your car as

there will be automatic vehicle system checks, firmware updates and data management services. And if a physical intervention is needed on the car, it would drive itself to the closest maintenance center while you are at work or sleeping at night.

You can have your car take care of your everyday life tasks without having to actually be physically traveling in the car. You can send your car pick-up your friend visiting you from the airport or train station, collect your suite from the cleaner or collect your grocery from the shop. You could even have your car go work for you while you are at work or at night and wake up in the morning richer with hundreds of dollars generated by your car giving transport services here and there while you are otherwise occupied. This sounds like science fiction, but these are possibilities with the state of the technology today.

The most significant global transformations of the 21st century will be the electrification of road transport and the final victory over a century of fossil-fuel transportation. The electric boom is now a reality. Are you ready to go get your slice of the pie?

"Intellectual growth should commence at birth and cease only at death."

- Albert Einstein

ELECTRIC BOOM!

GLOSSARY

This glossary of terms is gathering the most common technical jargon you will most likely encounter if you decide to dive deeper in the learning process of this very exiting industry:

A

All-electric range (AER)

This is a term used by California Environmental Protection Agency Air Resources Board[37] (CARB) which has legal meaning related to a requirement that a PHEV should be able to operate electrically until a specified set of conditions is no longer met.

Ampere

The ampere is that constant current which, if maintained in two straight parallel conductors of infinite length, of negligible circular cross-section, and placed 1 metre apart in vacuum, would produce between these conductors a force equal to 2×10^{-7} Newton per meter of length. The ampere unit is symbolized by "A".

Ampere-hour capacity

---

[37] The California Air Resources Board, also known as CARB or ARB, is the "clean air agency" in the government of California.

It is the quantity of electric charge measured in ampere-hours (Ah) that may be delivered by a cell or battery under specified conditions. One ampere-hour is the electric charge transferred by a steady current of one ampere for one hour. In EV applications, typical conditions involve a specific ambient temperature and a discharge time of 1 or 3 hours: in these cases the capacity is expressed as $C1$ or $C3$ (refer also to "Rated capacity", "Installed capacity", "Energy capacity").

B

Battery cell

A primary cell (single-use or "disposable") delivers electric current as the result of an electrochemical reaction that is not efficiently reversible, so the cell cannot be recharged efficiently. A secondary (rechargeable) cell is an electrolytic cell for generating electric energy, in which the cell, after being discharged, may be restored to a charged condition by sending a current through it in the direction opposite to that of the discharging current.

Battery module

A group of interconnected electrochemical cells in a series

and/or parallel arrangement, physically contained in an enclosure as a single unit, constituting a DC (direct-current) voltage source used to store electrical energy as chemical energy (charge) and to later convert chemical energy directly into electric energy (discharge). Electrochemical cells are electrically interconnected in an appropriate series/parallel arrangement to provide the module's required operating voltage and current levels. In common usage, the term "battery" is often also applied to a single cell. However, use of "battery cell" is recommended when discussing a single cell.

Battery pack

A completely functional system that includes battery modules, battery support systems, and battery specific controls. It may also be a combination of one or more battery modules, possibly with an added cooling system, and very likely with an added control system. A battery pack is the final assembly used to store and discharge electrical energy in a HEV, PHEV, or EV. (refer to Value Chain of Electric Car Battery Packs, p125)

Battery round-trip efficiency

The ratio of the electrical output of a battery pack on discharge to the electrical input required to restore it to the initial state of charge under specified conditions.

Battery Electric Vehicle (BEV)

An EV is defined as "any autonomous road vehicle

exclusively powered with an electric drive, and without any on-board electric generation capability: No gen-set onboard.

Battery State Of Charge (SOC)

It is the available capacity (or energy) in a battery expressed as a percentage of rated nominal capacity.

C

C rate

Discharge or charge current, in amperes, expressed in multiples of the rated capacity (refer to Rated Capacity definition below).

Capacitance

The ratio of the charge on one of the conductors of a capacitor (there being an equal and opposite charge on the other conductor) to the potential difference between the conductors. Capacitance is symbolized by "C".

Capacitor

A device which consists essentially of two conductors (such as parallel metal plates) insulated from each other by a dielectric (an insulator that may be polarized by an applied electric field). As part of an electric circuit, a capacitor introduces the capability of storing electrical energy, blocks the flow of direct current, and permits the flow of alternating current to a degree dependent on the capacitor's capacitance and the current frequency.

Charge / charging

The conversion of electrical energy, provided in the form of current from an external source, into chemical energy within a cell or battery. The (electrical) charge is also a basic property of elementary particles of matter.

Charge rate

The current at which a battery is charged (refer to C rate).

Charger

An energy converter for the electrical charging of a battery consisting of galvanic secondary elements.

Charge depletion

When a rechargeable electric energy storage system (RESS) on a PHEV, EV or extended-range EV is discharged.

Charging levels

Charging equipment is classified by the maximum amount of power in kilowatts provided to the battery. There are several levels of charging equipment. In North America, the standards are:

- Level 1: which is a 120-volt (V) alternating current (AC) plug. A full charge at Level 1 can take between 8 and 20 hours, depending on the battery capacity of the vehicle. Charging rate is approximately 1 kW.

- Level 2: which is a 240-volt AC plug and requires installation of home charging equipment. Level 2 charging can take between 3 and 8 hours, again depending on the battery capacity of the vehicle. Charging rates fall within a range of 3 kW to 20 kW.

- Direct Current (DC) fast charging: which is as high as 600 V, enables charging along heavy traffic corridors and at public stations. A DC fast charge can take less than 30 minutes to charge a battery to most of its capacity.

Controller

It is an element that restricts the flow of electric power to or from an electric motor or battery pack (module, cell). One purpose is for controlling torque and/or power output. Another may be maintenance of battery life, and/or temperature control.

Conventional vehicle

It is a vehicle powered by a conventional mechanical

drivetrain with internal combustion engine (ICE).

Current

It is the rate of transfer of electricity, meaning the amount of electric charge passing a point per unit time. The unit of measure is the ampere, which represents around 6.241×10^{18} electrons passing a given point each second.

Cut-off voltage

It is the cell or battery voltage at which the discharge is terminated. The cut-off voltage is specified by the cell manufacturer and is generally intended to limit the discharge rate.

Cycle

It is a sequence of a discharge followed by a charge, or alternatively a charge followed by a discharge, of a battery under specified conditions.

Cycle life

It is the number of cycles under specified conditions that are available from a battery before it fails to meet specified criteria regarding performance.

D

Depth of Discharge (DOD)

It is the percentage of electricity (usually in ampere-hours) that has been discharged from a battery relative to its rated nominal fully charged capacity (refer also to "Voltage efficiency", and "Watt-hour efficiency").

Direct current motor / DC motor

It is an electric motor that is energized by direct current to provide torque. There are several classes of direct current motors.

Discharge

It is the direct conversion of the chemical energy of a cell or battery into electrical energy and withdrawal of the electrical energy into a load.

Discharge rate

It is the rate, usually expressed in amperes, at which electrical current is taken from a battery cell, module, or pack (refer to "C rate").

Driving range

It is the maximum distance travelled by a vehicle, under specified conditions, before the "fuel tanks" need to be recharged. For a pure EV, it is the maximum distance travelled by a vehicle under specified conditions before the batteries need to be recharged. For a PHEV it will be the maximum distance achievable after emptying both the battery pack and fuel tank. For a conventional vehicle or HEV it will be the maximum distance achievable after emptying the fuel tank.

E

Electric drive system

It is the electric components that serves to drive the vehicle. This includes (a) electric motor(s), final control element(s), and controllers and software (control strategy).

Electric drivetrain (including electric drive system)

It is the electromechanical system between the vehicle energy source and the wheels. It includes controllers, motors, transmission, driveshaft, differential, axle shafts, final gearing, and wheels.

Electric Vehicle (EV)

An EV is defined as "any autonomous road vehicle exclusively powered with an electric drive, and without any on-board electric generation capability: No gen-set onboard.

Electric Vehicle Supply Equipment (EVSE)

It is the equipment that delivers electrical energy from an electricity source to charge an EV's battery. It communicates with the EV to ensure that an appropriate and safe flow of electricity is supplied. EVSE units are commonly referred to as "charging stations" or "charging points" and include the connectors, conductors, fittings and other associated equipment.

Electrochemical cell

It is the basic unit able to convert chemical energy directly into electric energy.

Energy capacity

It is the total number of watt-hours that can be withdrawn from a new cell or battery. The energy capacity of a given cell varies with temperature, rate, age, and cut-off voltage. This term is more common to system engineers than the battery industry, where the ampere-hour is the preferred unit and terminology.

Energy consumption

It is the energy consumed by a vehicle per unit distance (in km or miles) and, sometimes, also per unit weight (in tons). It may be expressed as kWh/km and also kWh/(ton-km). For EVs and PHEVs the electrical energy counted, expressed in AC kWh, is from the plug (charger input). Usually developed from tests of vehicles when driven over a "driving cycle" (a speed versus time requirement), with a specified passenger and/or luggage load. Standardized methods of estimating fuel consumption of PHEVs have not yet been developed.

Energy density

It is the ratio of energy available from a cell or battery to its volume in liters (Wh/L). The mass energy density in battery and EV industry is normally called specific energy (refer to "Specific energy").

Extended-range electric vehicle

Also known as a series PHEV, an extended-range electric vehicle is an "autonomous road vehicle" primarily using electric drive provided by a battery, but with an auxiliary on-board electrical energy generation unit and fuel supply used to extend the range of the vehicle once the battery charge has been depleted.

F

Fast Charging

Also known as "DC quick charging", fast charging stations provide a direct current of electricity to the vehicle's battery from an external charger. Charging times can range from half an hour to two hours for a full charge.

Fuel cell

It is an electrochemical cell that converts chemical energy directly into electric energy, as the result of an electrochemical reaction between reactants continuously supplied, while the reaction products are continuously removed. The most common reactants are hydrogen (fuel) and oxygen (also from the air).

Fuel cell vehicle (FCV)

It is a vehicle with an electric powertrain that uses the fuel cell as a source of the electricity to provide electric drive. FCVs may also include an electric storage system (ESS) and be HEVs or PHEVs. However, an ESS is not technically necessary in a FCV.

Fuel consumption

It is the energy consumed by a vehicle per unit distance (in km or miles) and, sometimes, also per unit weight (in tons). It may be expressed as kWh/km and also kWh/(ton-km). For EVs and PHEVs the electrical energy counted, expressed in AC kWh, is from the plug (charger input). Usually developed from tests of vehicles when driven over a "driving cycle" (a speed versus time requirement), with a specified passenger and/or luggage load. Standardized methods of estimating fuel consumption of PHEVs have not yet been developed.

Fuel economy

This is also referred to as fuel efficiency. For an EV it is the distance (in km or miles) travelled per unit energy from the plug, in kWh. For an internal combustion engine vehicle it represents the distance travelled per liter/gallon of fuel. It is the reciprocal of the energy per unit distance (the reciprocal of fuel consumption). Usually developed from tests of vehicles when driven over a "driving cycle" (a speed versus time requirement), with a specified passenger and/or luggage load. Standardized methods of estimating fuel economy of PHEVs have not yet been developed.

Full HEV

It is a full Hybrid Electric Vehicle that has the ability to operate all-electrically, generally at low average speeds. At high steady speeds such a HEV uses only the engine and mechanical drivetrain, with no electric assist. At intermediate average speeds with intermittent loads, both electric and mechanical drives frequently operate together. A PHEV can be developed based on a full HEV powertrain.

H

Hourly battery rate

It is the discharge rate of a cell or battery expressed in terms of the length of time during which a fully charged cell or battery can be discharged at a specific current before reaching a specified cut-off voltage. The hour-rate = C/i, where C is the rated capacity and i is the specified discharge current. For EVs, a 3-hour or a 1-hour discharge is preferred.

Hybrid electric vehicle (HEV) – Parallel configuration

A parallel hybrid is an HEV in which both an electric machine and engine can provide final propulsion power together or independently (refer to chapter 4 – Electrification levels).

Hybrid electric vehicle (HEV) – Series configuration

A series hybrid is a HEV in which only the electric machine can provide final propulsion power (refer to chapter 4 – Electrification levels). .

Hybrid vehicle

A vehicle with at least two different energy converters and two different energy storage systems for the purpose of vehicle propulsion.

I

Induction motor

It is an alternating-current (AC) motor in which the primary winding on one member (usually the stator) is connected to the power source, and the secondary winding on the other member (usually the rotor), carries only current induced by the magnetic field of the primary. The magnetic fields react against each other to produce a torque. This one of the simplest, reliable, and cheapest motors made.

Inductive charging

This involve the use of magnetic coupling devices instead of standard plugs in charging stations. This technology is actively pursued for EVs and opens new possibilities for charging.

Infrastructure

It is every part of the system except the vehicle itself that is necessary for its use. For PHEVs or EVs the infrastructure includes available fuel (electricity), power plants, transmission lines, distribution lines, access to parts, maintenance and service facilities, and an acceptable trade-in and resale market.

Installed capacity

It is the total number of ampere-hours that can be withdrawn from a new battery cell, module, or pack when discharged to the system-specified cut-off voltage at the HEV, PHEV, or EV design rate and temperature (i.e., discharge at the specified maximum DOD).

Internal combustion engine (ICE)

It is the historically most common means of converting fuel energy to mechanical power in conventional road vehicles. Air and fuel are compressed in cylinders and ignited intermittently. The resulting expansion of hot gases in the cylinders creates a reciprocal motion that is transferred to wheels via a driveshaft or shafts.

K

Kilowatt-hour (kWh)

One thousand (1000) watt-hours of energy, which also equals 1.341 horsepower- hours (or 1.35962 CVh).

L

Lithium ion (Li-ion)

The term "lithium-ion" refers to a family of battery chemistries. Li-ion chemistries commonly used today have come down significantly in cost and have increased gravimetric and volumetric energy density over the last 10-15 years, with progress accelerating in the last few years. Li-ion is seen as the coming enabling technology for PHEVs.

M

Motor, electric machine, generator

A motor is a label for an electric machine that most frequently converts electric energy into mechanical energy by utilising forces produced by magnetic fields on current-carrying conductors. Most electric machines can operate either as a motor or generator. When operating as a generator, the electric machine converts mechanical energy into electrical energy. In HEVs, PHEVs, and EVs, electric machines operate both in motoring and generating modes.

N

Neighborhood Electric Vehicle (NEV)

It is a vehicle defined in US Federal Regulations. NEVs are low-speed electric vehicles that have a maximum speed of 25 mph and can only be driven on roads with a maximum speed of 35 mph. Such vehicles have a much less stringent set of safety requirements than do other US light-duty vehicles.

New Energy Vehicle (NEV)

In China, NEVs are typically defined as plug-in hybrid electric vehicles (PHEVs), battery electric vehicles (BEVs), and fuel cell vehicles (FCVs).

New European Driving Cycle (NEDC)

This is a driving cycle consisting of four repeated ECE-15

driving cycles and an Extra-Urban driving cycle (EUDC). The NEDC is supposed to represent the typical usage of a car in Europe, and is used, among other things, to assess the emission levels of car engines. It is also referred to as MVEG cycle (Motor Vehicle Emissions Group).

Nickel cadmium (NiCd)

This was a common battery chemistry used in many EVs of the 1990s as well as in consumer electronics. It is no longer in common use because of restrictions put on hazardous substances, which include cadmium.

Nickel-metal hydride (NiMH)

Nickel metal hydride was also a common commercial battery chemistry in the 1990s for consumer electronics. In the late 1990s it became the battery of choice for HEVs. It has lower gravimetric and volumetric energy density than lithium-ion chemistries.

Nominal capacity

The total number of ampere-hours that can be withdrawn from a new cell or battery for a specified set of operating conditions including discharge rate (for EV, usually C1 or C3), temperature, initial state of charge, age, and cut-off voltage.

Nominal voltage

It is the characteristic operating voltage or rated voltage of a cell, battery, or connecting device.

Normal charging

It is the most common type and location for charging of an EV battery pack necessary to attain the state of maximum charge of electric energy. Also called slow or standard charge.

O

Opportunity charging

The use of a charger during periods of EV or PHEV inactivity to increase the charge of a partially discharged battery pack, thus increasing the range.

Overcharge

It is the forcing of current through a cell after all the active material has been converted to the charged state. In other words, charging is continued after 100% state of charge (SOC) is achieved.

P

Parallel battery pack

Term used to describe the interconnection of battery cells

and/or modules in which all the like terminals are connected together.

Parallel HEV

It is an HEV in which the engine can provide mechanical power and the battery electrical power simultaneously to drive the wheels.

Peak power (in kW)

It is the peak power attainable from a battery, electric machine, engine, or other part in the drive system used to accelerate a vehicle. For a battery this is based on short current pulse (per 10 seconds or less) at no less than a specified voltage at a given depth of discharge (DOD). For an electric machine, the limiting factor is heating of insulation of copper windings. Peak power of an engine is generally related to mechanical capabilities of metal parts at peak allowable revolutions per minute (rpm), also affected by heat. Generally, continuous power ratings are well below peak power ratings.

Plug-in hybrid electric vehicle (PHEV)

It is an HEV with a battery pack with a relatively large amount of kWh of storage capability, and with an ability to charge the battery by plugging a vehicle cable into the electricity grid. This allows more than two fuels to be used to provide the propulsion energy.

PHEVxk

A plug-in hybrid electric vehicle with "x" miles or kilometers of estimated charge depletion all electrically (CDE) range (also known as all-electric range, or AER). In this glossary, we suggest adding a small letter "k" to denote when the "x" values are in kilometres, or an "m" to denote when those values are in miles.

Power

It is the rate at which energy is released. For an EV, it determines acceleration capability. Power is generally measured in kilowatts.

Power density (volumetric)

It is the ratio of the power available from a battery to its volume in liters (W/L). The mass power density in battery and EV industry is normally called specific power (see "Specific power") or gravimetric power density.

R

Range

It is the maximum distance travelled by a vehicle, under

specified conditions, before the "fuel tanks" need to be recharged. For a pure EV, it is the maximum distance travelled by a vehicle under specified conditions before the batteries need to be recharged. For a PHEV it will be the maximum distance achievable after emptying both the battery pack and fuel tank. For a conventional vehicle or HEV it will be the maximum distance achievable after emptying the fuel tank.

Rare Earth Metals

It is a set of seventeen chemical elements in the periodic table, many of which are used in components in the drivetrains of hybrid and electric vehicles.

Rated capacity

It is the battery cell manufacturer's estimate of the total number of ampere-hours that can be withdrawn from a new cell for a specified discharge rate (for EV cells usually C1 or C3), temperature, and cut-off voltage.

Rechargeable electric energy storage system (RESS)

Battery packs, flywheels, and ultra-capacitors are examples of systems that could be repeatedly charged from the grid, with the charge later discharged in order to power an electric machine to move a vehicle.

Regenerative braking

It is a means of recharging the battery by using energy produced by braking the EV. With normal friction brakes, a certain amount of energy is lost in the form of heat created by friction from braking. With regenerative braking, the electric machines act as generators. They reduce the braking energy lost by returning it to the battery, resulting in improved range and improved lifespan of the friction brakes.

S

Self-discharge

The loss of useful electricity previously stored in a battery cell due to internal chemical action (local action).

Series HEV

A series hybrid is a HEV in which only the electric machine can provide final propulsion power (refer to chapter 4 – Electrification levels).

Slow Charging

It is the most common type of charging that provides alternating current to the vehicle's battery from an external charger. Charging times can ranges from 4 to 12 hours for a full charge.

Smart charging

It is the use of computerized charging devices that constantly monitor the battery so that charging is at the optimum rate and the battery life is prolonged.

Specific energy, or gravimetric energy density (of a battery)

The energy density of a battery expressed in watt-hours per kilogram.

Specific power, or gravimetric power density (of a battery)

The rate at which a battery can dispense power measured in watts per kilogram.

Start-stop

It is the lowest level of electrification of a powertrain, involving a slightly larger (higher kW) electric machine and battery than for starting alone, providing an ability to stop the engine when the vehicle is stopped and save fuel that would have been consumed at engine idle.

Start-stop + regeneration (and electric launch)

It is a technology package that can also be called "minimal" or "soft" hybridization. According to the International Society of Automotive Engineers (SAE), a hybrid must provide propulsion power. If a start-stop system includes regeneration and electric launch, it is a hybrid, according to the SAE definition. If it does not, it is not a hybrid.

State of charge (SOC)

It is the available capacity (or energy) in a battery expressed as a percentage of rated nominal capacity.

T

Three-phase Controller

An electronic circuit for controlling the output frequency and power from a 3-phase inverter.

Total Cost of Ownership (TCO)

It is the purchase price of a vehicle plus the costs of operation during the time period it is owned. Costs include depreciation costs, fuel costs, insurance, financing, maintenance, and taxes.

U

Useable capacity

It is the number of ampere-hours (or kilowatt-hours) that can be withdrawn from a battery pack installed in an EV, taking into account decisions on control strategy designed to extend battery pack life or achieve vehicle performance goals. Useable capacity is a smaller number than nominal capacity, usually 80% of the total battery capacity..

V

Volt

It is a unit of potential difference or electromotive force in the International System units, equal to the potential difference between two points for which one Coulomb of electricity will do 1 Joule of work in going from one point to the other. The volt unit is symbolised by "V".

Voltage efficiency

The ratio of the average voltage during discharge to the average voltage during recharge under specified conditions of charge and discharge.

W

Watt-hour efficiency

It is the ratio of the watt-hours delivered on discharge of a battery to the watt-hours needed to restore it to its original state under specified conditions of charge and discharge.

Watt-hours per kilometer

It is the energy consumption per kilometer or mile at a particular speed and condition of driving. It is a convenient overall measure of a vehicle's energy efficiency. Watt-hour efficiency = Ampere-hour efficiency x voltage efficiency.

Z

Zero emission vehicle (ZEV)

A vehicle that has no regulated emissions out of the tailpipe. Under California Air Resources Board (CARB) regulations, either an EV or a FCV is also a ZEV.

ELECTRIC BOOM!

TABLE OF ILLUTRATIONS

ABOUT THE AUTHOR

Mohammed Belbaraka is an author, engineer, speaker, and entrepreneur. He has over fifteen years of experience in the industry. He has a Master's degree in engineering and has a strong, successful background in design, production, system engineering and program management.

He is currently working for the most prestigious OEMs across North America and Europe. Mohammed Belbaraka has become an expert in the field of transport electrification.

ELECTRIC BOOM!

DISCLAIMER

The facts set out in this book are obtained from sources which we believe to be reliable. However, we accept no legal liability of any kind for this book contents, nor any information contained therein nor conclusions drawn from it by any party. The author and publisher accept no responsibility for the consequences of any actions resulting from the information in this book.

NOTES

MOHAMMED BELBARAKA

ELECTRIC BOOM!

MOHAMMED BELBARAKA

ELECTRIC BOOM!